Nathalia Souza Bezerra
Adna Cristina B. de Sousa

Biological control of leaf-cutting ants

AF625305

Nathalia Souza Bezerra
Adna Cristina B. de Sousa

Biological control of leaf-cutting ants

Efficiency of entomopathogenic fungi on leaf-cutting ants (Hymenoptera: Formicidae)

ScienciaScripts

Imprint

Cover image: www.ingimage.com

This book is a translation from the original published under ISBN 978-613-9-63620-4.

Publisher:
Sciencia Scripts
is a trademark of
Dodo Books Indian Ocean Ltd. and OmniScriptum S.R.L publishing group

120 High Road, East Finchley, London, N2 9ED, United Kingdom
Str. Armeneasca 28/1, office 1, Chisinau MD-2012, Republic of Moldova, Europe
Printed at: see last page
ISBN: 978-620-8-10827-4

CONTENTS

Bezerra, Nathalia Souza. Efficiency of entomopathogenic fungi on leaf-cutting ants (Hymenoptera: Formicidae). 2018. 93 p. Dissertation (Master's Degree) - Postgraduate Program in Biotechnology. Biotechnology Center, Federal University of Paraiba (UFPB), *Campus* I, Joao Pessoa, Paraiba, Brazil.

SUMMARY

The leaf-cutting ants of the genera *Atta* and *Acromyrmex* are considered important pests of the Neotropics. Currently, the main control method employed is the use of chemical insecticides. However, there is growing concern about the negative effects of frequent insecticide applications on non-target organisms, human health and the environment. Entomopathogenic fungi can naturally infect host insects via direct penetration of the cuticle and have been studied worldwide as promising biological control agents. That said, the aim of this work is to evaluate the efficiency of isolates of *Beauveria bassiana*, *Beauveria brongniartii*, *Metarhizium anisopliae* var. *anisopliae*, *Paecilomyces* sp. and *Aspergillus* sp. on leaf-cutting ants in the laboratory. In addition to analyzing the effect of Fipronil on the vegetative growth and conidiogenesis of the fungal isolates. For each species of fungus, an experiment was carried out with 90 specimens of leaf-cutting ants, corresponding to three replicates of 10 ants for each treatment (Treatment 1θru as control = autoclaved distilled water + Tween 80 0.01 %; Treatment 2 = suspension of 1.0 x 104 conidia.mL^{-1} + Tween 80 0.01 %; Treatment 3 = suspension of 1.0 x 10^8 conidia.mL^{-1} + *Tween* 80 0.01 %). The ants were transferred individually to Petri dishes containing filter paper with 1 mL of the conidia suspension and cotton balls moistened with a 10% honey solution. The plates were kept at a temperature of 25 ± 2 °C and relative humidity of 70 ± 10 % and evaluated every 24 hours for 10 days to observe the extrusion of the fungi and confirm that the insects had been killed by the pathogens. To determine the effect of the insecticide on the vegetative growth and conidiogenesis of the fungal isolates, 0.8 g/L of the product and 0.3 g/L of penicillin were added to the Sabouraud-Dextrose Agar culture medium. After 10 days, the diameter of each colony was measured using a ruler and the conidia were counted in a Neubauer chamber. Fipronil was then rated for its in vitro toxicity to the isolates. The results of the bioassay showed that the two concentrations of conidia used (1.0 x 104 and 1.0 x 108 conidia.mL^{-1}) were capable of ensuring the infection and progress of the disease in leaf-cutting ants, with the average mortality of the ants varying between 4.80 ± 2.10 and 9.99 ± 0.00 and the TL50 varying between 6.01 and 10.22 days. In relation to conidial concentrations, it was observed that, in some cases, increasing the concentration contributed to higher mortality in the insects. Among all the isolates, Aspergillus sp. had the highest average mortality (9.99 ± 0.10 and 9.99 ± 0.00) and the lowest TL_{50} (6.47 and 6.01 days) for both conidia concentrations. The fungal isolates treated with Fipronil showed a reduction in vegetative growth. However, the insecticide did not significantly compromise their conidiogenesis. It can therefore be concluded that the bioassay described in this study allowed the discovery of isolates with great potential to be evaluated in the field for the biological control of leaf-cutting ants, contributing to the development of more effective and less environmentally damaging alternatives for the control of these insect pests and that Fipronil, at the concentration tested, proved to be "compatible" for the fungal isolates *in the in vitro* compatibility test.

Keywords: Biotechnology, leaf-cutting ants, biological control.

1. INTRODUCTION

Leafcutter ants, genera Atta and Acromyrmex, cause damage to agricultural and forestry crops by cutting plant material, mostly leaves, and using it as a substrate for growing a symbiotic fungus, which is the main food for the larvae, queen and adults. Often, due to the size of the anthill, a large volume of leaves is needed to meet the population's food demand. It is estimated that leafcutter ants are capable of collecting between 22 and 940 kg (dry weight) of plant matter per colony per year. Due to their defoliating potential, they are considered one of the five most important pests in Latin America, accounting for around 75% of the costs and time spent on pest control (VILELA, 1986 apud CANTARELLI et al., 2008; LOECK et al., 2001; RICCI et al., 2005 apud TEJEDA et al., 2017; DANS et al., 2009; SILVA et al., 2013a; ARAGÓN-GARCIA et al., 2016).

The control of leaf-cutting ants has been carried out predominantly through the use of chemical insecticides such as Fipronil and Sulfluramid (BRASIL, 2010 apud ORTIZ et al., 2017). This control strategy has presented major difficulties due to environmental contamination, low specificity, the possibility of pest resurgence, the selection of resistant insects and potential risks to human and animal health. Given this situation, there has been growing scientific interest in recent years in finding and implementing methods that are less aggressive to the environment, animals and humans (LEMUS et al., 2008; GUERRERO, 2013; BARBOSA et al., 2015; MARQUES et al., 2016).

Biological control with entomopathogens is seen as a viable, safe and environmentally friendly alternative for controlling insect pests (CHAVEZ et al., 2016). Entomopathogenic organisms are those that show specificity and infect insects. This group includes viruses, bacteria, protozoa, nematodes and fungi (GUERRA et al., 2001). Among these entomopathogenic agents, we highlight the filamentous fungi Metarhizium anisopliae, Beauveria bassiana, Beauveria brongniartii, Paecilomyces sp. and Aspergillus sp. These fungi naturally infect more than 300 species of insects from various orders, including important pests of agriculture, livestock and public health (SILVA, 2017). In view of these aspects, better knowledge of the dynamics, from specificity to fungal colonization in a target population, related to the horizontal transmission of fungal propagules, will contribute to more effective control of the targeted pest.

Basic studies that characterize the virulence of these fungi are important for the selection of a new agent to be exploited in the biological control of various insect pests, including leaf-cutting ants. The prospect of the real pathogenicity of entomopathogenic fungi against leaf-cutting ants and the development of new technologies for the application of these biological products could contribute to more effective control, to minimizing the damage caused and to the possible eradication of leaf-cutting ants in agricultural and forestry crops. Therefore, our objective was to evaluate the action of five different entomopathogenic fungal isolates on leaf-cutting ants in the laboratory, with the aim of contributing to the development of more effective and less environmentally damaging alternatives for controlling this insect pest.

2. THEORETICAL BASIS

2.1. Leaf-cutting ants

Ants of the genera *Atta* (saùvas) and *Acromyrmex* (quenquéns) (Formicidae, Myrmicinae, Attini) are popularly known as "leaf-cutting ants" (RIBEIRO et al., 2012; ARAÙJO et al., 2015; DIEHL et al., 2017). The saùvas are characterized by the presence of three pairs of dorsal spines, while the quenquéns have four or more pairs of dorsal spines (TRAVAGLINI et al., 2017).

Leafcutter ants are found in most Neotropical *habitats*. They have a habit of cutting plant matter such as leaves, flowers and fruits from different plant species, which are pre-processed into small pieces and transported to the anthill, where they are used as a substrate for growing fungi (ABREU; DELABIE, 1986; FARJI-BRENER; MEDINA, 2000; HAEDER et al., 2009; FRANCO et al., 2013; LEAL et al., 2014; MATTE et al., 2016; DIEHL et al., 2017). Ants prefer the younger, softer parts of plants, as well as foliage rich in water, nitrogen and phosphorus, with little fiber and manganese (RANDO; FORTI, 2005; RODRiGUEZ et al., 2008).

The fungus-substrate combination is called a "fungus garden" (DORNELAS et al., 2016). Fungus gardens are located in underground chambers in the anthill (RODRIGUES et al., 2014). The cultivated fungus is a Basidiomycete and belongs to the *Leucoagaricus* genus, developing properly in conditions of high humidity and temperatures between 20 °C and 30 °C (DELLA LUCIA et al., 2014; ROMER et al., 2017).

There is a mutualistic relationship between ants and fungi that originated over 50 million years ago, in which one cannot survive without the other (SANTOS; CAZETTA, 2016). The fungus has enzymes capable of degrading plant biomass, making nutrients accessible that were previously unavailable (AYLWARD et al., 2013; VIGUERAS et al., 2017). Part of the products released during this process is consumed by the fungus, meeting its energy and nutrient demands, as well as enabling the production of specialized structures called *gongylidia,* which serve as food for the ants. The gongylidia contain a mixture of carbohydrates, amino acids, proteins, Iipids and vitamins (SUEN et al., 2011; VIGUERAS et al., 2017). The ant larvae and the queen feed on gongylidia and fungal hyphae, while the workers feed in a mixed way, with the fungus representing around 9% of their nutritional needs, the rest being obtained from additional sugars present in the plant sap, which are consumed during leaf cutting. The importance of fungi as food for the larvae is due to the fact that they contain the nutrients necessary for the proper development of the larvae (ROCKWOOD; HUBBELL, 1987; LUNA, 1996; POULSEN et al., 2002; KOST et al., 2007; MASIULIONIS, 2013; SANTOS; CAZETTA, 2016). The ants, in turn, nourish the fungus with leaf fragments, provide a favorable environment for its growth and protect it from competitors and invading organisms. In addition, they ensure its spread when winged females carry small portions of the fungus to new colonies (BOT et al., 2002; SANTOS et al., 2004; DUARTE et al., 2014).

The cultivated fungus is essential for the survival and architecture of the ant colony (ZANETTI et al., 2014). For this reason, ants have developed different strategies to help maintain the integrity of the fungus. These include meticulous cleaning of nests and leaf fragments used as substrates, in order to remove fungal spores and bacteria; intense self-grooming and allogrooming behavior (cleaning and licking between different individuals in the colony); and secretion of antimicrobial substances produced by their metapleural glands (ARISMENDY; PERALTA, 2002; BOT et al., 2002; SOUZA et al., 2003; RODRIGUES et al., 2008; SCHOENIAN et al., 2011; RIBEIRO et al., 2012; DELLA LUCIA et al., 2014). In addition, ants maintain an association with filamentous bacteria (Actinomycetes) present in their integuments, which are capable of producing secondary metabolites that protect both ants and gardens against pathogenic microorganisms. These metabolites are known to inhibit the growth of a parasitic fungus belonging to the genus *Escovopsis,* which destroys fungal food crops. In turn, the growth of these bacteria is apparently maintained through the secretion of nutrients by specialized cuticular glands. The predominant actinomycetes belong to the genus *Pseudonocardia* (DELLA-LUCIA et al., 1995; JÙNIOR et al., 2001; CURRIE et al., 2003; CAFARO et al., 2011; DÂNGELO, 2011; RODRIGUES et al., 2014; GÓMEZ, 2017).

A great diversity of symbiotic microorganisms can be found in fungus gardens, which can perform diverse functions, such as defense against pathogens and promoting the growth of the cultivated fungus through nitrogen fixation (SCHOENIAN et al., 2011). Some bacterial communities apparently also help the garden fungus to degrade the plant substrate (SUEN et al., 2011).

The nests of leafcutter ants are formed by a system of underground tunnels and chambers, which can reach depths as great as seven meters and are capable of housing millions of individuals (KLEINEIDAM et al., 2001; MEYER et al., 2013). During the construction of the chambers, the workers remove the soil and deposit it on the surface, forming a mound of loose earth (**Figure 01**) (MOREIRA et al., 2007). In addition to the chambers that are used to grow the fungus, there are additional chambers where dead ants, exhausted fungus, debris and organic remains that are removed from the fungus gardens are deposited (FARJI-BRENER; MEDINA, 2000; MOUTINHO et al., 2003; SOARES et al., 2006).

Figure 01. Leaf-cutter ant nest.

Source: Author (2017).

Leafcutter ants are eusocial insects, with a high level of social organization in the form of a caste system, where individuals perform specific functions (TEIXEIRA et al., 2014). The permanent castes are made up of queens and workers. The queen is the fertile female who has lost her wings after the nuptial flight. She has a larger body size than the others and can live for many years, in some cases twenty years or more. During the pairing with the male, the queen stores enough sperm to fertilize eggs for the rest of her life. There is only one queen per colony. The workers, on the other hand, are sterile and polymorphic and include the soldiers, responsible for defending the colony, the gardeners who work in the fungus gardens, the cutters and loaders, responsible for cutting and transporting the plant material, the nurses, responsible for looking after the eggs, larvae and pupae and the garbage collectors, responsible for placing the waste generated in separate chambers to avoid possible contamination. The temporary castes are made up of virgin queens and winged males that appear during the reproductive period. At certain times of the year, the winged males and females leave their original colonies to found new ones, and are responsible for perpetuating the species. Dispersal for mating takes place under favorable conditions of rainfall, temperature, humidity and wind speed. Males die soon after copulation. Before the nuptial flight, the females take a piece of the fungus with them in their infrabuccal sac, which will allow them to form a new fungal crop (ARAÙJO et al., 2003; FORTI et al., 2004; LERMA et al., 2006; ALVES-FAVA et al., 2011; CAMARGO et al., 2013; ESTRADA; SERMENO-CHICAS, 2013; GÓMEZ, 2017).

2.1.1. Leafcutter ants: ecological importance

When in balance with their environment, leaf-cutting ants are very important from an ecological point of view, acting at various levels within an ecosystem. They are called "ecosystem engineers" because of their ability to modify the physical, chemical and biological structures of their own environment. Ants promote soil aeration by removing large amounts of soil for the construction of anthill chambers and tunnels, substantially improve soil penetrability, disseminate seeds, accelerate the cycling of bioelements and increase the availability of nutrients in the soil through the accumulation of nutrient-rich plant material. Some studies report that the level of nutrients available in the soil where ants nest is higher than in areas adjacent to anthills. In addition, the waste chambers present in leaf-cutter ant nests can serve as a *habitat* for some species (HAINES, 1978 apud VITTAR, 2008; DAUBER; WOLTERS, 2000 apud SANTOS, 2016; GRUBER; VALDIX, 2003; CORRÊA et al., 2010; SALINAS; DiAZ, 2011; ALVIRA, 2014; ALVIRA; ACOSTA, 2015; TRAVAGLINI et al., 2017).

2.1.2. Leafcutter ants: economic importance

Leaf-cutting ants are one of the main agricultural and forestry pests in the Neotropics, due to their abundance, wide distribution and the economic losses they cause by partially or totally defoliating agriculturally important crops to feed the mutualistic fungus they cultivate. They are considered the dominant herbivores of the Neotropics, since they consume much more vegetation than any other group of animals of comparable taxonomic diversity, carrying out their damaging action all year round. The damage they cause varies according to the age of the plant and the intensity of the attack. In certain regions, certain crops can be discarded because the ants won't let them thrive. The damage caused by ants to agriculture has been estimated at around 1 billion dollars a year. In addition to the direct damage caused to plants, leafcutter ant nests can reduce access to crops and also damage roads (HOLDOBLER; WILSON, 1990 apud MASON et al, 2017; DELLA LUCIA, 2003; LOECK et al., 2003;

SANTOS et al., 2007b; AUBAD, 2010 apud CELIS, 2013; CASTILHO et al., 2010; FILHO et al., 2011b; ESTRADA et al., 2013; FARDER-GOMES et al., 2017; OLIVEIRA et al., 2017).

2.1.3. Control of leaf-cutting ants

Cutter ants can be controlled using biological, cultural, mechanical and chemical methods (BOARETTO; FORTI, 1997).

2.1.3.1. Biological methods

In general, biological control can be defined as the rational use of predators, parasitoids and pathogens to regulate the population density of pests, preventing them from reaching levels capable of causing economic damage (BALE et al., 2008; SIMONATO et al., 2014).

Predators, parasitoids and pathogens are called "natural enemies" and reduce the population density and reproductive potential of weed species, thus reducing the damage they cause to crops (LEÓN, 2005).

The successful use of this method depends on knowledge of the biology and ecology of the pests, as well as the natural enemies (VAN DEN BOSCH et al., 1982 apud GASSEN; TAMBASCO, 1983).

Predators can be defined as free-living organisms that attack, kill and feed on many individuals during their life cycle. They are usually larger than their prey (BASTOS; TORRES, 2003; OLIVEIRA et al., 2006). An example of a predator are insects belonging to the Coccinellidae family, popularly known as ladybugs. Ladybugs have been used in the biological control of primary and secondary agricultural pests, including aphids, mealybugs, whiteflies and defoliator caterpillars, among others (GUERREIRO, 2004).

Parasitoids are organisms whose immature stages develop at the expense of a host, causing its death or making it unable to generate offspring (BRECHELT, 2004; BADII; ABREU, 2006; CRUZ et al., 2011; ROSSI, 2012). Parasitoids of agricultural pests are mainly dipterans or hymenopterans (GONZALES, 2017). They generally parasitize a single host during their life cycle. The host can be the egg, larva or pupa of pests (BASTOS; TORRES, 2003; CRUZ et al., 2011). An example of a parasitoid is the wasp *Trichogramma* sp. The adult female oviposits inside the eggs of the pest. After some time, larvae hatch which feed on the contents of the host's eggs and develop into adult wasps which continue the cycle (CRUZ, 2007).

Entomopathogens are microorganisms that cause diseases in insects. This group is very diverse and includes bacteria, viruses, fungi, nematodes and protozoa. Each of these subgroups varies in its way of infecting and in its pathogenic mechanism (BADII; ABREU, 2006; CANTÙ-RUIZ et al., 2017). Some pathogens have a broad host spectrum, while others prefer certain insect species (CANTÙ-RUIZ et al., 2017).

2.1.3.2. Cultural methods

An example of this type of control is the revolving of the soil (plowing and harrowing) to eliminate the initial nests of leaf-cutter ants (MIRANDA; ARAÙJO, 2003; FILHO et al., 2013). Another example is the use of attractive plants near the crops, which serve as alternative food for the ants or act as "traps", preventing them from attacking the crops (FILHO et al., 2011a).

2.1.3.3. Mechanical methods

Mechanical control is generally little used (MORESSI et al., 2007). One example is the mechanical removal of nests using shovels. During this procedure, an internal crisis is created in the colony, which can lead to the death of the queen and part of the brood or even make the nest more susceptible to predators or diseases (MONTOYA-CORREA et al., 2007). Another example of this type of control is the use of physical barriers, such as inverted plastic cones, adhesive gel or plastic strips covered in grease which are fixed to the trunks of plants and prevent leaf-cutting ants from gaining access to the aerial part, where the leaves, flowers and fruit are located (NAVA et al., 2009; ALMEIDA et al., 2013).

2.1.3.4. Chemical methods

It is currently the most widely used control method, with the use of granulated baits, post-drying, post-soluble and thermal spraying standing out (LINK et al., 2000; TORRES et al., 2013; ARAÙJO et al., 2015).

Toxic baits are widely used due to their low cost and ease of application (FILHO et al., 2007). They consist of an attractive substrate, usually dehydrated citrus pulp, impregnated with insecticide (PERRI et al., 2017). The right insecticide must have a delayed action, be lethal at low concentrations and act through ingestion (FORTI et al., 2003). The baits are deposited near the ant trails and nest entrances. The efficiency of the baits depends on the attractive substrate and the toxic active ingredient used, the size of the granules, as well as the ants' ability to locate them and transport them into the colony. If the baits are not sufficiently attractive, they may not be transported by the ants (FILHO; OLIVEIRA, 2002; RAMOS et al., 2006; DELLA LUCIA et al., 2014; GANDRA et al., 2016).

Thermonebulization is a process suitable for large areas and is carried out using appropriate devices called thermonebulizers (CRUZ et al., 1984). This method consists of atomizing a formicide in kerosene, diesel oil or mineral oil using heat, resulting in a toxic smoke that is applied inside the anthill. Most of the time, this is done until it is saturated, i.e. until the smoke applied to one of the anthill's openings escapes through the others (SILVA et al., 2008; BUENO, 2013). It can be used at any time of the year, on dry or waterlogged land, and is highly efficient. However, there is a greater risk of contamination of the environment, as well as of the operator himself due to exposure to the product (SANTOS et al., 2007a; ZANETTI et al., 2008; TRAVAGLINI et al., 2017).

The advantages of chemical control include practicality, efficiency, speed and low cost (LENTEREN; WOETS, 1988; GHINI; BETTIOL, 2000; LIMA et al., 2011). However, there are some unfavorable aspects associated with chemical methods, such as contamination of the environment, selection of resistant populations, poisoning of humans and animals, low specificity, elimination of natural predators, the fact that chemical insecticides can be incorporated into the food chain, among others, making it necessary to search for control methods that reconcile efficiency, economy and safety (CANTARELLI et al., 2005; ALMEIDA et al., 2007; CASTILHO et al., 2010; ALMEIDA et al., 2013).

2.2. Biological pest control with entomopathogenic fungi

Entomopathogenic fungi are capable of naturally regulating insect populations (DiAZ et al., 2006). It is estimated that around 80% of the diseases that affect insects are caused by fungi (BADII; ABREU, 2006).

Entomopathogenic fungi were the first biological agents to be used in pest control (DELGADO; MURCIA-ORDONEZ, 2011). Louis Pasteur and Agostino Bassi were among the first scientists to recognize the potential of using microorganisms to control insect pests and, in the mid-1870s, this notion of microbial control was expressed by several prominent entomologists. It was Metchnikoff who demonstrated in practice that an entomopathogenic fungus could infect an insect pest in the soil and applied the fungus for insect control (TANADA; KAYA, 1993).

Entomopathogenic fungi have a broad spectrum of action and can colonize various species of flies, grasshoppers, aphids, caterpillars, beetles, etc., and are used as biological control agents for economically important insect species. The soil is their main reservoir (VELASQUEZ et al., 2007; VALICENTE, 2009; SCHAPOVALOFF et al., 2015).

Metarhizium, Beauveria, Paecilomyces, Aspergillus, Verticillium and Aschersonia are among the main genera of entomopathogenic fungi and act as a regulatory mechanism for insect populations in ecosystems (MONZÓN, 2001; CASTILLO et al., 2012). These include Metarhizium anisopliae and *Beauveria bassiana.* These are the two best-studied entomopathogenic fungal species and have shown great potential as microbial control agents, being widely used to control pests in various crops such as coffee, sugar cane and bananas, among others (MOINO JR. et al., 2002; POTRICH et al., 2009).

Metarhizium anisopliae (Metsch.) Sorokin is pathogenic to more than 204 species of insects, belonging to the orders Orthoptera, Lepidoptera, Hymenoptera, Coleoptera, among others, causing what is known as "green muscardine" in insects. This fungus is widely distributed throughout the world and was isolated by Metchnikoff in 1879 from the beetle *Anisoplia austriaca.* Two varieties have been described, M. anisopliae var. anisopliae and M. anisopliae var. majus, which differ in the size of their conidia (GUERRERO et al., 2000; GUTIÉRREZ; SALDARRIAGA, 2004; BAUTISTA-GALVEZ; GONZÂLEZ- CORTES, 2005; DAMIN et al., 2011).

The conidiophores of M. *anisopliae* are simple. The conidia are oval to cylindrical and truncated at both ends. In each conidiophore, a chain of basipetalous conidia is formed, which grow dense and adhered to each other, forming prismatic masses of columns. The conidia are clavate or cylindrical. In the *anisopliae* variety, the conidia are 3.5 to 9 μm long, while in the *majus* variety, the conidia are 9 to 18 μm long. The colonies of M. *anisopliae* vary in shape and, at the start of growth, are white in color. Mature colonies turn yellowish, olive green or dark green (ALVES, 1998; GUERRERO et al., 2000; GUTIÉRREZ; SALDARRIAGA, 2004; GOMEZ et al., 2008; STERLING et al., 2011).

The process of the fungus penetrating the insect's integument, colonization, generalized infection and death of the host usually takes place between three and ten days after contact (MELO et al., 2007). Hosts affected by this fungus become mummified and are covered in a powdery layer formed by the agglomeration of conidia, which can vary in color from light to dark green, greyish or even whitish with green dots (ALVES, 1998; ALBUQUERQUE et al., 2005; OTTATI-DE-LIMA, 2007).

In Brazil, *M. anisopliae* has been studied on various pests, including *Mahanarva posticata* (sugarcane leafhopper), *Diatraea saccharalis* (sugarcane borer), *Hypothenemus hampei* (coffee borer), *Nezara viridula* (soybean greenbug), among others (DAMIN et al., 2011).

Beauveria bassiana Balls. (Vuill.) was discovered by Agostino Bassi in 1835 and is a cosmopolitan species that naturally inhabits the soil (XIAO et al., 2012; MASCARIN; JARONSKI, 2016). It is known for causing "white muscardine" in insects (PAULI et al., 2011). It has a variety of hosts, including insect vector species, agricultural pests and invasive insects belonging to the orders Coleoptera, Diptera, Lepidoptera and Hemiptera (GHIKAS et al., 2010; PEDRINI et al., 2013; HUANES-CARRANZA; WILSON-KRUGG, 2016).

It is one of the most widely used fungi in pest control programs, due to its efficiency and easy multiplication. In the United States, B. *bassiana* has been used to control adults of *Plectrodera scalator*, achieving mortality rates of over 60% (LEITE et al., 2011). In China, it is estimated that around one million hectares are treated each year with *B. bassiana* in order to control insects, including the caterpillar *Dendrolimus punctatus* (BARBOZA et al., 2011; XIAO et al., 2012). After its release, B. *bassiana* can persist in the soil for periods of time ranging from 15 days to 1 year (LEITE et al., 2011).

The *B. bassiana* colony has a cottony to powdery appearance and is usually white in color, which can vary to yellowish-white. The conidiophores are simple and irregular. The phialides have a globose or bulky base and are thinner at the top, forming a curved, *zigzag-shaped* sterigma. The conidia are hyaline and smooth, round to ovoid and unicellular (GOMEZ et al., 2008; GONZÂLEZ-CASTILLO et al., 2012; RODRIGUES et al., 2016).

The mycelium and conidia of B. *bassiana* cover the body or joint spaces of attacked insects with a white coating (CASTILLO et al., 2012).

Beauveria brongniartii (Sacc.) Petch (Hypocreales: Cordycipitaceae) is the second most common species of the genus *Beauveria* and is found especially in the soil (GHIKAS et al., 2010; MAYERHOFER et al., 2015; SABBOUR; ABDEL- RAHEEM, 2015).

B. brongniartii is a natural pathogen of the Coleoptera: Scarabaeidae (PIATTI et al., 1998). It has been used as an agent in the biological control of *Melolontha melolontha* L.

(Coleoptera: Scarabaeidae), whose larvae and adults cause considerable damage to a variety of crops in Central Europe by feeding on the roots and leaves of plants and are considered pests in agriculture (CRAVANZOLA et al., 1997; KESSLER et al., 2004; DOLCI et al., 2006; FATU et al., 2015; SIERPINSKA et al., 2015).

Several biological products based on the fungus B. *brongniartii* have been marketed in Europe, such as Melocont®-Pilzgerste, *Beauveria brongniartii* Myzel and *Beauveria* Schweizer, Engerlingspilz (FATU et al., 2015).

The cologne is cottony to powdery. Initially, it is white in color and, as it ages, it can take on a "pale yellow" hue. The conidiophores are sparse and rarely in clusters. The conidiogenous cells are shaped like sub-globose or enlarged "bottles" and the conidia are ellipsoid (ALVES, 1998; FLORES, 2003; CANEDO; AMES, 2004; SVEDESE, 2012).

The genus *Aspergillus* includes more than 200 species, some of which are capable of producing secondary metabolites with great entomopathogenic and entomotoxigenic potential (MORAES et al., 2001; LOPES et al., 2004). Species such as *A. flavus* and *A. parasiticus*, for example, have been used to control mosquitoes *(Culex quinquefasciatus)* due to the larvicidal effect of their metabolites. The conidia of *A. clavatus* have also been shown to be effective against mosquitoes (SEYE et al., 2014).

Macroscopically, *Aspergillus* colonies show a white color in the initial stage of maturation, which can later evolve to yellow, brown, black or green, depending on the species (CARVALHO, 2013). The vegetative mycelium is composed of septate hyphae that branch dichotomously at 45° angles (LOPES et al., 2004). The conidiophore of *Aspergillus* is formed by the foot-cell, usually in the shape of an "L" or "T", the stem (stipe) and the vesicle. On top of the vesicle are the conidiogenous cells, phialides, responsible for promoting asexual reproduction through the production of conidia (phialoconidia), which in turn are hyaline, globose, unicellular and arranged in chains (ABARCA, 2000; XAVIER et al., 2008; SILVA, 2009; COSTA et al., 2014; REIS; ROCHA, 2014).

If the conidial heads have the phialides inserted directly into the vesicle, they are classified as "uniseriate". When there is a second layer of cells (metulas) connecting the phialides to the vesicle, the conidial heads are classified as "biseriate" (LIRA, 2014).

The genus *Paecilomyces* was described in 1907 and includes several entomopathogenic species, including *P. farinosus* and *P. fumosoroseus, which* are responsible for the disease "yellow muscardine" (SANTOS, 2003; LOPES, 2007).

P. farinosus (Holm ex S. F. Gray) Brown & Smith is a cosmopolitan fungus that parasitizes various insects belonging to the orders Lepidoptera, Diptera, Coleoptera, Hymenoptera and Hemiptera (LOPES, 2007).

P. fumosoroseus (Wize) Brown & Smith is one of the most important species in the genus. It is a pathogen that has a broad host range and a wide geographical distribution, being pathogenic to more than 40 species of insects, including lepidopterans, coleopterans, dipterans, homopterans and isopterans. It can be isolated from soil or insects (PÉREZ et al., 2003; SANTOS; NÛNEZ, 2003; JACKSON et al., 2004; GONZÂLEZ-CASTILLO et al., 2012). P. *fumosoroseus* has been successfully used as a biocontrol agent for whitefly (Hemiptera: Aleyrodidae) and other insect pests. In addition, two types of propagules can be used for *P. fumosoroseus* formulations, blastospores (produced in liquid medium) and conidia (produced in solid medium) (VEGA et al., 1999; CASTELLANOS-MOGUEL et al., 2008).

The colony can be whitish, yellow, reddish or pinkish in color. The conidiophores of *Paecilomyces* sp. can be simple or in synema, verticillate and with long spinules. The conidia originate at the end of the phialides, forming long chains, and can be elliptical to fusiform, hyaline or pigmented (ALVES, 1998; LOPES, 2007; FAIA, 2011).

Biological control with entomopathogenic fungi is a very promising technology, in which very effective isolates are used and the main fungal propagules used in this practice are conidia, blastospores or mycelium (PARI et al., 2015; LEITE et al., 2003 apud SCHAMNE, 2010).

2.3. Biological mechanisms of action of entomopathogenic fungi on their hosts

There are four main routes of infection: through the integument, the mouth cavity, the spiracles and other external openings in the insect. The exact site of infection depends on the fungus, the insect and the existing conditions such as humidity and temperature (MADELIN, 1963; TÉLLEZ-JURADO et al., 2009). In general, the mechanism of action of entomopathogenic fungi includes the following stages: adhesion, germination, formation of appendages, formation of penetration structures, penetration, colonization and reproduction (GARCiA et al., 2008). Adhesion, germination and penetration are highly important processes for the start of the pathogen cycle and occur during the first few hours after contact between the host and the pathogen (LECUONA et al., 2005). The duration of the different stages of the fungal cycle depends on the species and also on the environmental conditions during infection (CASTILLO et al., 2012).

Initially, viable conidia adhere to the insect's cuticle. This conidium-cuticle interaction involves the action of glycoproteins, extracellular enzymes, electrostatic forces and hydrophobic forces. In addition, many entomopathogenic fungi have conidia coated with a mucilaginous matrix that favors their adhesion to the cuticle (EVANS, 1993; CASSIANO et al., 2008; GARCiA et al., 2008).

Germination occurs when the conidium finds favorable conditions (humidity, temperature and nutritional requirements in the cuticle) and begins with its swelling (DiAZ et al., 2006; MARANHÂO; MARANHÂO, 2008; 2009). After the conidium swells, the germ tube begins to form, which traces and recognizes the surface of the insect. Subsequently, the germ tube undergoes morphological changes, forming the appressorium. The appressoria are swellings in the apical region of the germ tubes, where there is a high level of metabolic activity due to the production of enzymes. The appendix also acts to anchor the conidia and exerts pressure on the interior of the insect. From the appressorium, the penetration clamp is formed, a tapered structure that is in contact with the cuticle and initiates the penetration process (TIAGO; FURLANETO, 2003; DiAZ et al., 2006; RODRiGUEZ-DEL-BOSQUE; ARREDONDO-BERNAL, 2007; CARRILLO- RAYAS; BLANCO-LABRA, 2009; ZHANG et al., 2010; VERISSIMO, 2015).

Physical and chemical mechanisms are involved in penetration. The first consists of the mechanical pressure exerted by the penetration structures, which breaks through membranous or poorly sclerotized areas. The second consists of the production and release of hydrolytic enzymes, mainly proteases, lipases and chitinases. The combination of these factors makes the cuticle fragile, allowing the fungus to cross it and gain access to the general cavity of the insect's body (MONZÓN, 2001; MARANHÂO; MARANHÂO, 2008; 2009; GAJARDO et al., 2012; GIMENES et al., 2014). Once inside the hemocele, the hyphae enlarge and branch out inside the host's tissue, producing blastospores, unicellular or multicellular structures that lose their cell wall, but have a thin fibrillar layer on the plasma membrane, and are responsible for nourishing the fungus through the degradation of carbon sources present in the insect's hemolymph (MARANHÂO; MARANHÂO, 2008; 2009; RAMiREZ et al., 2014; VERISSIMO, 2015). Upon consuming the nutrients, the fungus begins the process of mycelial growth, spreading within the host through the circulation (COVA et al., 2009; AGALI et al., 2017).

The attack on the host insect involves both direct infection by the fungus and the action of secondary metabolites produced by it, some of which are toxic to insects. Toxic metabolites include extracellular enzymes, proteins and low molecular weight compounds. Among the toxins produced by entomopathogenic fungi are the destruxins produced by M. anisopliae and the beauvericin produced by B. bassiana (MARANHÂO; MARANHÂO, 2008; 2009; FAN et al., 2013a; FAN et al., 2013b; MORA et al., 2016). Some fungi apparently have no toxins and kill the insect by consuming all its nutrients (RAMiREZ et al., 2014).

Once infected with the fungus, insects can show symptoms such as dark spots, convulsions, lack of coordination, paralysis and altered behavior. The insect stops feeding, becomes weak and visually disoriented. They die within a few days due to a combination of factors such as nutrient depletion, fungal toxins, physical obstruction of circulation, invasion of organs, blockage of the digestive system and other physical damage caused by the fungus' vegetative growth (DIODATO, 1992; POULSEN et al., 2006; MARANHÂO; MARANHÂO, 2008; 2009; TÉLLEZ-JURADO et al., 2009; GAJARDO et al., 2012; FAN et al., 2013a).

Colonization begins in the haemocele and then continues to the rest of the body, such as the nervous system, digestive system, muscles, trachea, Malpighian tubules and fat bodies (MARTiNEZ, 2014). The fungus colonizes practically all the insect's organs and tissues (SANCHEZ et al., 2001). The hyphae eventually emerge from the cadavers and these become covered with a mycelial mass (SANCHEZ et al., 2001; AGALI et al., 2017). If conditions are not favorable, the fungus remains inside the insect, protected by the integument, and can survive for some time until external conditions are suitable (RAMÎREZ et al., 2014).

The mycelial mass contains conidiophores which, through their conidiogenic regions, give rise to new conidia (SANCHEZ et al., 2001; SCHAPOVALOFF et al., 2015). The attacked insects become mummified and start to show a characteristic color for each species of fungus. For example, white in the case of B. bassiana and olive green in the case of M. anisopliae (PASTOR, 2013). Conidia can be spread either by abiotic agents such as wind or rain or when sick individuals come into contact with healthy individuals, continuing the biological cycle of the fungus (SANCHEZ et al., 2001; BADII; ABREU, 2006; SCHAPOVALOFF et al., 2015).

2.4. Advantages and limitations of using entomopathogenic fungi in the biological control of insect pests

Among the main advantages of biological control of insect pests with entomopathogenic fungi is the fact that these entomopathogens are generally quite specific and do not affect humans. In addition, they can be easily cultivated en masse (HANSON; HILJE, 1993). Other advantages include the fact that, unlike other microorganisms such as bacteria and viruses, entomopathogenic fungi do not need to be ingested to be effective. In addition, they act on a range of organisms and are also capable of infecting hosts at different stages of their development, such as eggs, larvae, pupae and adults, they have a low impact on the environment and on beneficial fauna, they have great genetic variability and have a high capacity to prevent hosts from developing resistance. Furthermore, the way in which entomopathogenic fungi reproduce gives them greater persistence over time, since the dead insect becomes a new source of inoculum (FRANCE et al., 2000; QUESADA-MORAGA et al., 2006; MELO et al., 2007; DELGADO; MÛRCIA-ORDONEZ, 2011; BHAN et al., 2013; GUTIÉRREZ et al., 2016; 2017).

The main limitations of biological control with entomopathogenic fungi include the lack of knowledge about the principles of the method, the lack of qualified personnel and the lack of economic support (GUÉDEZ et al., 2008). Other limitations are related to the fact that fungi are usually slow-acting pathogens, and their ability to infect a target insect is influenced by biotic and abiotic factors (CARRILLO-RAYAS; BLANCO-LABRA, 2009; ECKARD et al., 2017). Factors such as the behavior, physiological conditions and age of the host, as well as high temperatures, low humidity and UV radiation, can compromise the growth and development of the fungus on the host (MARANHÂO; MARANHÂO, 2008; 2009; BOTERO et al., 2009; YÂNEZ; FRANCE, 2010). Among the abiotic factors, UV radiation is one of the most important, as it can cause lethal damage to DNA and mutations, causing delayed germination or the inactivation of conidia, thus reducing bioinsecticidal activity (OTTATI-DE-LIMA et al., 2010; ZORZETTI et al., 2014).

2.5. Integrated pest management

The term Integrated Pest Management (IPM) was coined in the mid-1960s and emerged as a reaction by researchers to the excessive use of insecticides to combat pests (BORJA, 2003; GONZAGA et al., 2009). It can be defined as "a decision-based process involving the coordinated use of multiple tactics to control all classes of pests (insects, pathogens, weeds and vertebrates) in an ecologically and economically safe manner" (EHLER, 2006). In addition to the ecological and economic aspects, IPM has to take into account the sociological aspect, i.e. any measure to be adopted must seek the well-being of society, since it will consume the agricultural products produced (CARVALHO; BARCELLOS, 2012). In practice, IPM consists of mastering some basic knowledge about the correct identification and behavior of pests, their population dynamics, their ecology, their seasonal fluctuation and natural mortality (effect of climate and natural enemies), among others, and monitoring the target pest is the starting point for choosing the most appropriate control method (FILHO et al., 2004; SALVADORI; PEREIRA, 2006; LAURENTIS, 2017). IPM represents a significant step forward, as its main objective is the minimal use of agrochemicals, helping to mitigate environmental contamination problems and reduce residue rates in the final product, thus improving the quality of life of both producers and consumers (FILHO et al., 2004).

The combination of chemical pesticides and entomopathogenic fungi could be an alternative in which the agrochemical, in sublethal doses, is expected to act as a stressor on the target population, making the insects more susceptible and increasing the efficiency of the mycopathogen. This strategy aims to achieve greater control efficiency, taking into account both economic and ecological results. However, it is necessary to make sure that there is compatibility, since certain chemical pesticides can interfere with the performance of the microorganism. In view of this, synthetic insecticides recommended by specialized entities should be used for this type of control. These recommendations take into account certain parameters, such as the efficiency and toxicity of the products, as well as their effect on the natural enemies of the pests and the risks of their application (STERN et al., 1959; HUFFAKER, 1969; HOFFMANN-CAMPO et al., 2000; JÛNIOR et al., 2000; FILHO et al., 2001; DALZOTO; UHRY, 2009; BORGES; NOVA, 2011; FILHO; RODRIGUES, 2015).

2.6. Fipronil

Fipronil and Sulfluramid are examples of synthetic insecticides that have been used to control leaf-cutting ants (TORRES et al., 2013).

Fipronil is an insecticide from the phenyl pyrazole chemical group (BRITTO et al., 2016). It was the first phenyl pyrazole insecticide to be introduced in pest control (IKEDA et al., 2004). It was discovered in 1987 by Rhône-Poulenc Agro and presented for the first time at the

Brighton Conference in 1992 (GANT et al., 1998).

It is considered extremely toxic to insects (SIMAS et al., 2001). It acts on the central nervous system of insects, more specifically on gamma-aminobutyric acid (GABA) receptors associated with chloride channels, blocking the passage of chloride ions. This eliminates the normal inhibition of nerve impulses, causing an increase in neural activity and, consequently, paralysis and death of the organism (IKEDA et al., 2004; LOURENÇO et al., 2012; TOFOLO et al., 2015; CHAGURI, 2016). It is known in more than 70 countries and is used to control pests in more than 100 different crops (OKUMURA, 2009). In Brazil, it was first used in 2000 on plantations in the Northeast and is currently being used to control termites, beetles, caterpillars and borers on crops such as cotton, potatoes, corn, soybeans, among others. Fipronil is also used as an antiparasitic agent in domestic animals to control fleas and ticks (MASUTTI; MERMUT, 2007 apud MOREIRA et al., 2012; SILVA et al., 2012). In the United States, it is registered for various uses such as the control of fleas, ticks, termites, ants, among others (GROSMAN et al., 2002). However, Fipronil can cause undesirable toxic effects for non-target species such as fish and aquatic invertebrates, as well as for the environment, potentially contaminating soil and water sources (UDO, 2012; FRATTI; TORRES, 2014).

3. OBJECTIVES

3.1. General objective:

J To evaluate the action of *Aspergillus* sp., *Beauveria bassiana, Beauveria brongniartii, Metarhizium anisopliae* var. *anisopliae* and *Paecilomyces* sp. on leaf-cutting ants and the effect of the insecticide Fipronil on the growth and conidiogenesis of these fungi.

3.2. Specific objectives:

J Evaluate the action of *Aspergillus* sp., *B. bassiana, B. brongniartii, M. anisopliae* var. *anisopliae* and *Paecilomyces* sp. on leaf-cutting ants, in the laboratory, using two concentrations of conidia, 1.0 x 10^4 conidia.mL^{-1} and 1.0 x 108 conidia.mL^{-1} .

J Determine the average mortality of leaf-cutting ants exposed to different concentrations of conidia from isolates of entomopathogenic fungi.

J Determine the median lethal time (TL_{50}) for each fungal isolate at both conidia concentrations.

J To evaluate the effect of the insecticide Fipronil on the vegetative growth and sporulation (conidiogenesis) of *Aspergillus* sp., B. *bassiana, B. brongniartii, M. anisopliae* var. *anisopliae* and *Paecilomyces* sp. *in vitro.*

J To classify the insecticide Fipronil *in terms* of its toxicity to the five different isolates of entomopathogenic fungi *in vitro.*

4. MATERIAL AND METHODS

The experiments were conducted at the Laboratory of Molecular Genetics and Plant Biotechnology (LAGEMOL), Biotechnology Center, Federal University of Paraiba (UFPB), *Campus* I, Joao Pessoa, Paraiba.

4.1. Origin of the insects and ticks: the specimens of leaf-cutting ants (Hymenoptera: Formicidae) were collected from anthills in the Atlantic Forest reserve located on *Campus* I of the Federal University of Paraiba in Joao Pessoa, Paraiba, in 2017. The insects were collected using entomological tweezers and placed in glass jars. They were then taken to the Molecular Genetics and Plant Biotechnology Laboratory (LAGEMOL/CBIOTEC/UFPB) for the bioassay. The ticks (adult females of *Boophilus microplus)* used to reinvigorate the fungal isolates were collected in the municipality of Bayeux, Paraiba, in 2017.

4.2. Origin of the fungal isolates: five fungal isolates specified in **Table 01** were used.

Table 01: Origin of the five entomopathogenic fungal isolates of *Beauveria bassiana*, *Beauveria brongniartii*, *Metarhizium anisopliae* var. *anisopliae*, *Paecilomyces* sp. and *Aspergillus* sp.

Fungal species	Insulation substrate/Origin
-*Beauveria bassiana* URM 2915	*Nezara viridula* /ˑ CENARGEM / PR - 1987
-*Beauveria brongniartii* URM 6504	Agroforestry system soil / Abreu e Lima / PE - 2011
*-Metarhizium *anisopliae* var. *anisopliae* URM 4920	*Mahanarva posticata* / Usina Serra Grande / Maceió /AL - 2005
¥*Paecilomyces* sp. TP08	Diesel oil / Gas station /Joao Pessoa / PB - 2014
Aspergillus sp.	Leaf-cutter ant / Atlantic Forest Reserve / Joao Pessoa / PB - 2017

- CENARGEM - National Agricultural Center for Genetic Resources. * Lineage used in the biological control of the sugarcane leafhopper in the North, Northeast and Southeast. -Strains provided by the URM Mycoteca of the Federal University of Pernambuco. ¥ Lineage provided by the Environmental Microbiology Laboratory of the Federal University of Paraiba (LAMA/CBIOTEC/UFPB). Source: Author (2017).

4.3. Culture medium and maintenance of fungal cultures: Agar-Sabouraud-Dextrose 2% (5 g/L meat peptone, 20 g/L glucose, 5 g/L casein peptone and 15 g/L bacteriological agar. pH 5.6 ± 0.1 at 25 °C) (Isofar, Duque de Caxias, RJ, Brazil). The methodology described by Guimaraes (2016) was used, adapted. The samples were transferred to test tubes containing 2% Agar-Sabouraud-Dextrose medium, where the cultures were kept at room temperature for 10 to 15 days and then refrigerated at 4 °C.

4.4. Microscopic examination of the fungal isolates: the methodology described by Lopes et al. (2008) was used, adapted. Inocula of the fungal cultures were transferred to Petri dishes containing Agar-Sabouraud-Dextrose medium and covered with previously flambéed Iaminulas. The plates were kept at room temperature and the coverslips with adhered mycelium were removed successively, 24 - 48 - 96 - 120 hours after the start of the experiment, placed inverted on a sterile slide, stained with lactophenol, identified and analyzed under an optical microscope.

4.5. Reinvigoration of the fungal isolates: the methodology used was adapted from the work of Ito et al. (2007). Before carrying out the bioassay, the isolates were passed on to adult female B. microplus ticks. To do this, the ticks were treated with a 0.5% sodium hypochlorite solution and then washed with autoclaved distilled water. They were then immersed for a few seconds in a suspension of fungal conidia + 0.01% Tween 80. The negative control was made using only autoclaved distilled water and Tween 80 0.01 %. The treated ticks were then transferred to Petri dishes containing filter paper moistened with autoclaved distilled water. The plates were kept at a temperature of 25 ± 2 °C and relative humidity of 70 ± 10 % for 10 days and the fungi were re-isolated from the tick cadavers.

4.6. Preparation of conidial suspensions: the methodology used was adapted from the work of Guimaraes (2016). Aspergillus sp., B. bassiana, B. brongniartii, M. anisopliae var. anisopliae and Paecilomyces sp. were inoculated onto 2% Agar-Sabouraud-Dextrose culture medium. The plates were kept at 25 ± 2 °C for a period of 10 to 15 days for growth and conidiogenesis. After this period, the conidia were collected by scraping the surface of the colony for the bioassay. To prepare the conidial suspensions, autoclaved distilled water and 0.01 % Tween 80 were used. The concentration of conidia was then estimated in a Neubauer chamber and the suspensions were standardized at 1.0 x 10^4 conidia.mL^{-1} and 1.0 x 108 conidia.mL^{-1} for each fungal isolate.

4.7. Pathogenicity test: the methodology used for the pathogenicity test was adapted from the work of Santos et al. (2007b). For each species of fungus, an experiment was carried out with 90 specimens of leaf-cutting ants, corresponding to three replicates of 10 ants for each treatment (T1Grupo cont lroe = autoclaved distilled water + Tween 80 0.01 %; T2= suspension of 1.0 x 104 conidia.mL^{-1} + Tween 80 0.01 %; T3= suspension of 1.0 x 10^8 conidia.mL^{-1} + Tween 80 0.01 %). The ants were transferred individually to Petri dishes containing filter paper with 1 mL of the conidia suspension and cotton balls moistened with a 10% honey solution. The plates were kept at a temperature of 25 ± 2 °C and relative humidity of 70 ± 10 %, and were evaluated every 24 hours for 10 days to observe the extrusion of the fungi and confirm that the insects had been killed by the pathogens.

4.8. Evaluation of the effect of the insecticide Fipronil on the vegetative growth and sporulation of entomopathogenic fungal isolates: the methodology used was based on the work of Rashid et al. (2012). The 2% Agar-Sabouraud-Dextrose culture medium was autoclaved at 121 °C for approximately 20 minutes. After cooling for a few minutes at room temperature, the antibiotic penicillin at a concentration of 0.3 g/L and 0.8 g/L of the insecticide Fipronil were added to the medium. The latter was previously filtered using 0.22 µm diameter filters in order to remove contaminants. The flask containing the Agar-Sabouraud-Dextrose medium, penicillin and Fipronil was manually shaken for one minute to homogenize the components and then poured into sterile Petri dishes. After the medium had solidified, the plates were kept in a refrigerator at 4 °C for 24 hours and then inoculated with the fungal isolates using a platinum loop (one spot per plate). Negative controls were made by growing the isolates on Agar-Sabouraud-Dextrose medium containing only penicillin, without Fipronil. The experiments were carried out in triplicate and the plates were kept at 25 ± 2 °C for 10 days. At the end, the diameter of the fungal colonies was obtained using a millimeter ruler and the number of conidia produced per colony was counted in a Neubauer chamber under an optical microscope.

4.9. Classification of the insecticide Fipronil in terms **of toxicity to different isolates of entomopathogenic fungi:** for the classification of Fipronil in terms of toxicity to isolates of *Aspergillus* sp., B. *bassiana, B. brongniartii, M. anisopliae* var. *anisopliae* and *Paecilomyces* sp. in *vitro,* the model proposed by Alves et al. (1998) was used. The percentage values are calculated in relation to the control (100%), as shown in **Equation 1**.

$$T = \frac{20(CV) + 80(ESP)}{100}$$

Equation 1

Where:

T = corrected value of vegetative growth and sporulation for product classification.

CV = percentage of vegetative growth compared to the control.

ESP = percentage of sporulation compared to the control.

Based on the T values, the chemical was classified as *very toxic* (0 to 30), *toxic* (31 to 45), *moderately toxic* (46 to 60) or *compatible* (> 60) (ALVES et al., 1998).

4.10. Statistical analysis

4.10.1. Pathogenicity test with leaf-cutting ants

The experiment was carried out using a completely randomized experimental design, where the data was statistically analyzed for variance (F test) and the means were compared with each other using the Tukey test at a 5% probability level, using the Sisvar computer program (FERREIRA, 2003).

Data on confirmed mortality was submitted to *Probit* analysis to obtain TL values$_{50}$ (in days) (FINNEY, 1971).

4.10.2. Fipronil test

The data were statistically analyzed for variance (ANOVA), followed by Duncan's multiple range test, where $P < 0.05$ (DUNCAN, 1955).

5. Results and discussion

5.1. Micrographs of fungal isolates

The vegetative and reproductive structures present in the different fungal isolates were analyzed using the Iaminula culture technique (**Figure 02**). For all the isolates, the germination of conidia and the formation of hyaline branched mycelium were evident within the first 24 hours. After 48 hours, mycelial differentiation was observed and morphological differences were identified between the vegetative and reproductive structures of the different isolates.

Figure 02. Laminate culture of the *Aspergillus* sp. isolate. **(A)** Laminate culture on Agar-Sabouraud-Dextrose medium. **(B)** Laminate with adhered mycelium stained with lactophenol to visualize fungal structures under an optical microscope.

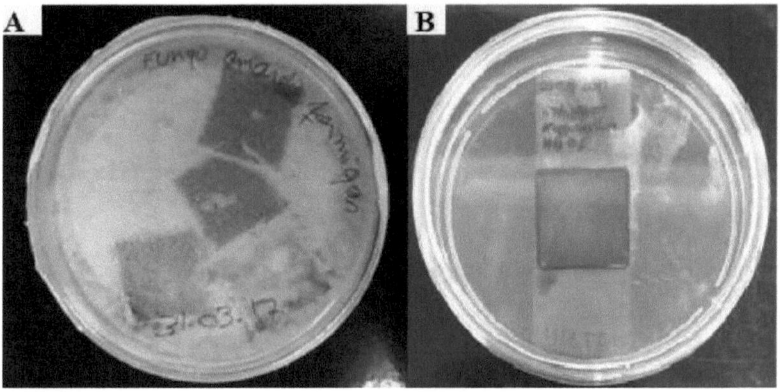

Source: Author (2017).

From the micrographs of *Aspergillus* sp. (**Figure 03A**), it was possible to observe the presence of a typical structure called the "aspergillus head", formed by the vesicle, the conidiogenous cells and the conidia, which is supported by a stem that starts from a foot cell. The conidia, in turn, are arranged in chains (ALVES, 2012; CARVALHO, 2013).

B. bassiana revealed septate, hyaline hyphae and the presence of conidiophores forming dense clusters. The conidia are unicellular, globose or subglobose and are arranged on the stems of the phialides. These, in turn, have *zig-zag* ends (SILVA, 2013; RODRIGUES et al., 2016) (**Figure 03B**).

C. *brongniartii* showed mycelial differentiation formed by hyaline, smooth, septate and uninucleate hyphae. The presence of spinules arranged on the main hyphae was noted, with an elongated to cylindrical shape. Conidiophores were scarce, rarely in clusters, where uninucleate and ellipsoid conidia were arranged (ALVES, 1998; SILVA, 2013) (**Figure 03C**).

M. anisopliae var. *anisopliae* showed a swelling at the end of one of the hyphae called an "appressorium", i.e. a specialized structure that originates from the germ tube or a hypha and which is essential to the process of penetrating the host (**Figure 03D**). In addition to hyaline, septate hyphae, anastomoses, simple conidiophores, elongated and uninucleate conidia (BONALDO; PASCHOLATI, 2007).

Paecilomyces sp. formed simple conidiophores and verticillate branches. The phialides had a swollen basal portion and an elongated portion, forming a distinctive neck, which produced and supported ellipsoid, unicellular conidia grouped in long chains (**Figure 03E**) (GÓMEZ et al., 2008; GONZALEZ-CASTILLO et al., 2012).

The correct identification of fungal isolates is important, especially in the case of entomopathogenic fungi, since these are used in the biological control of insect pests in agriculture (LUNA-ALVES LIMA, 1989 apud SILVA, 2013). Entomopathogenic fungi are generally identified by microscopic observation of the reproductive structures (HUMBER, 2012; LAZO, 2012).

The genera *Metarhizium*, *Beauveria*, *Paecilomyces* and *Aspergillus* belong to the subdivision Deuteromycotina, class Hyphomycetes (TANADA; KAYA, 1993).

In filamentous fungi, the asexual state is called "anamorph" and asexual reproduction occurs through the production of conidia. Spores can be formed inside unicellular structures (endospores) or externally (exospores) (VERiSSIMO, 2007; TAVARES, 2012).

Figure 03. Micrographs of the isolates of entomopathogenic fungi. **(A)** Conidia and conidiophore - *Aspergillus* sp. **(B)** Conidia and conidiophore - *Beauveria bassiana.* **(C)** *Conidia* and conidiophore - *Beauveria brongniartii.* **(D)** *Apressorium - Metarhizium anisopliae* var. *anisopliae.* **(E)** Conidia in chain - *Paecilomyces* sp.

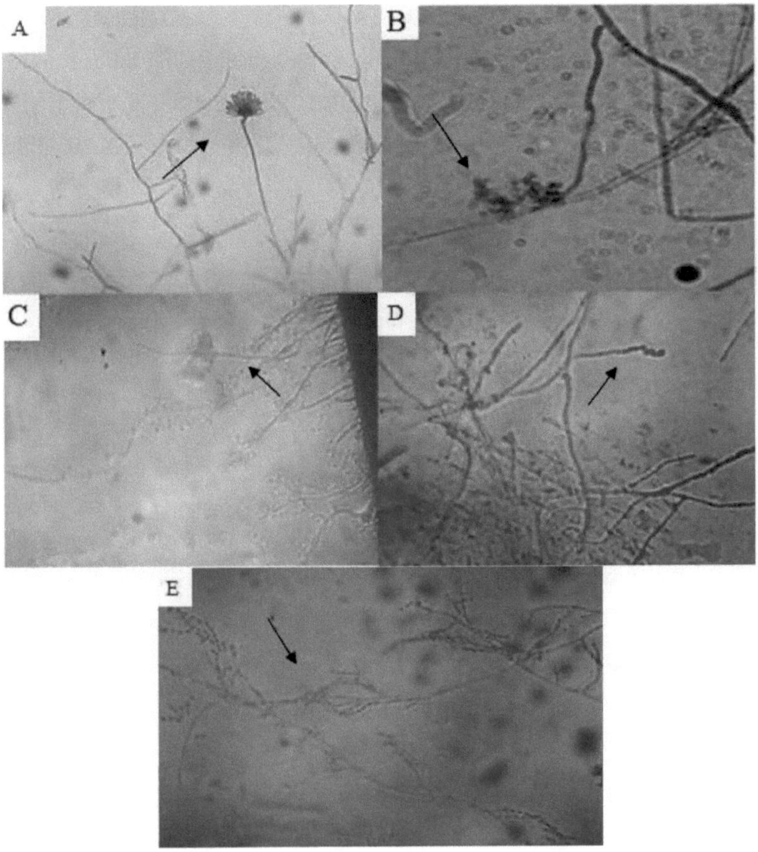

Source: Author (2017).

The different morphological types of conidia found in these genera are classified as exospores, as they are formed at the ends of the conidiophores (BERGAMIN FILHO et al., 1995 apud SILVA, 2013). Conidia are asexual exospores that act as infective agents and their main function is to perpetuate the species. They originate from special cells, called "conidiogenic cells", present in differentiated structures called "conidiophores". They are disseminated to distant locations by the action of winds or rain and germinate in a favorable environment (GUSMÂO; GRANDI, 1997; JUNIOR et al., 2005; FELIPINI; PIERO, 2009; TAVARES, 2012; SOUZA, 2013; GOMES, 2017).

These five isolates are considered to be eukaryotic fungi. This means that the reproductive structures are only formed in certain portions of the stalk (ALEXOPOULOS; MIMS, 1979 apud JORGE, 1998). The remaining part corresponds to the vegetative form (GUIMARÂES, 2016). The vegetative mycelium is responsible for penetrating and digesting the substrate, absorbing nutrients and acting as a support element (NEDER, 1992; BORGES, 1999). The reproductive mycelium, in turn, is responsible for the production of spores and generally extends into the air (NEDER, 1992).

The differentiation of somatic and reproductive structures, the arrangement of conidiophores and the shape of conidia are important characteristics for distinguishing species within different genera (ALVES, 1998; ELENA; PAPAS, 2002).

5.2. Reinvigorating fungal isolates

Four fungal isolates were passed on to adults of the B. *microplus* tick to activate their pathogenicity (**Figure 04**).

Figure 04. Invigoration of different entomopathogenic fungal isolates by passage through adults of the tick Boophilus microplus. **(A)** Beauveria bassiana. **(B)** Beauveria brongniartii. **(C)** Metarhizium anisopliae var. anisopliae. **(D)** *Paecilomyces* sp. Incubation period - 10 days (T= 25 ± 2 °C; Relative humidity 70 ± 10 %).

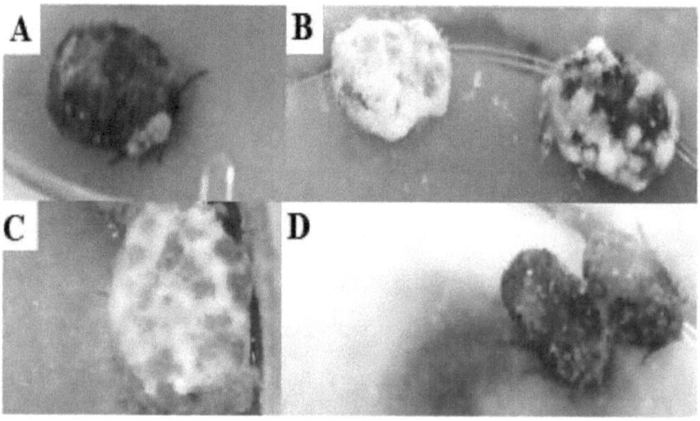

Source: Author (2017).

In the case of B. *brongniartii* and M. *anisopliae* var. *anisopliae*, it was possible to observe the complete mummification of the ticks at the end of the ten-day experiment (**Figure 04B** and **Figure 04C**, respectively). On the other hand, *B. bassiana* and *Paecilomyces* sp. showed satisfactory colonization, with hyphae exteriorizing over the tick's cuticle (**Figure 04A** and **Figure 04D**, respectively).

Fungi that are cultivated in the laboratory on artificial culture media can, over time, reduce their virulence when they are repeated. In *M. anisopliae*, for example, there was a reduction in virulence and changes in morphology after eight repicings (ALVES, 1998; ALVES; FARIA, 2010). Storage time in Mycotheque Culture Collections and exposure to high temperatures are also factors that contribute to reducing the capacity for infection in entomopathogenic fungi (LOPES et al., 2016). It is therefore recommended to reinvigorate the isolates by passing the pathogen over a host in order to activate its pathogenicity and enhance its control action on the pest (LOPES et al., 2008; MARiN; PARDEY, 2008; SANTOS et al., 2009).

The *Aspergillus* sp. isolate, on the other hand, did not undergo this invigoration process, as it was isolated directly from leaf-cutter ant specimens, after which it was repeated a few times in artificial culture media in the laboratory (**Figure 05A**).

After invigoration, the fungal species were re-isolated from the tick and cultivated on Agar-Sabouraud-Dextrose medium for the bioassay (**Figure 05**). Macroscopically, the colonies of all the isolates showed species-specific characteristics. *Aspergillus* sp. formed a rounded colony with a white vegetative part and a mass of yellow-green conidia on the front of the colony and an orange pigmentation on the back (**Figure 05A**). The *B. bassiana* colony had a velvety appearance and a whitish pigmentation on the front of the colony and cream on the back (**Figure 05B**). *B. brongniartii* showed a light cream-colored mealy appearance on the front of the colony and a darker pigmentation on the back (**Figure 05C**). *M. anisopliae* var. *anisopliae* showed a cottony appearance with greenish conidia on the front of the colony and a yellowish pigmentation on the back (**Figure 05D**). *Paecilomyces* sp. formed a mealy colony with a light brown color on the front of the colony and a yellowish pigmentation on the back (**Figure 05E**).

Figure 05. Macroscopic appearance of fungal colonies. **(A)** *Aspergillus* sp. **(A1)** Front. **(A2)** Back. **(B)** *Beauveria bassiana*. **(B1)** Front. **(B2)** Back. **(C)** *Beauveria brongniartii*. **(C1)** Front. **(C2)** Back. **(D)** *Metarhizium anisopliae* var. *anisopliae*. **(D1)** Front. **(D2)** Back. **(E)** *Paecilomyces* sp. **(E1)** Front. **(E2)** Back. Incubation period - 10 to 15 days (Agar-Sabouraud-Dextrose medium; T= 25 ± 2 °C; Relative humidity 70 ± 10 %).

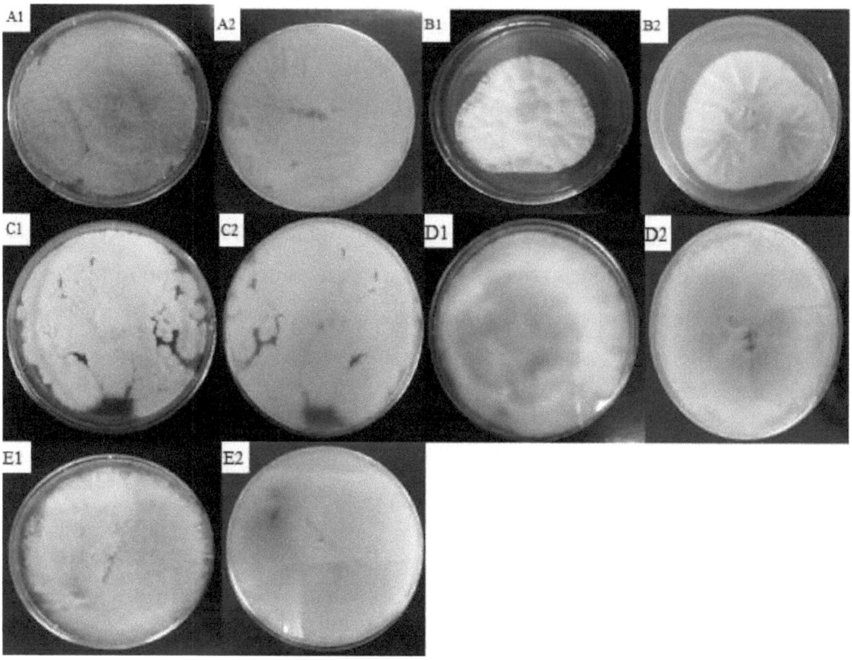

SOURCE: Author (2017).

5.3. Pathogenicity test with leaf-cutting ants

At the start of the bioassay, the ants were agitated, walking quickly in different directions (**Figure 06A**). Over time, they began to exhibit other types of behavior, such as cutting the filter paper on the plates (**Figure 06B**). This behavior, in particular, may be related to the very biology of leaf-cutting ants, which, when in their natural environment, have the habit of cutting leaves to feed the fungus they cultivate (HAEDER et al., 2009).

Figure 06. Bioassay with leaf-cutting ants. **(A)** Start of the bioassay with leaf-cutting ants. **(B)** Highlight of the behavior exhibited by leaf-cutting ants during the laboratory test.

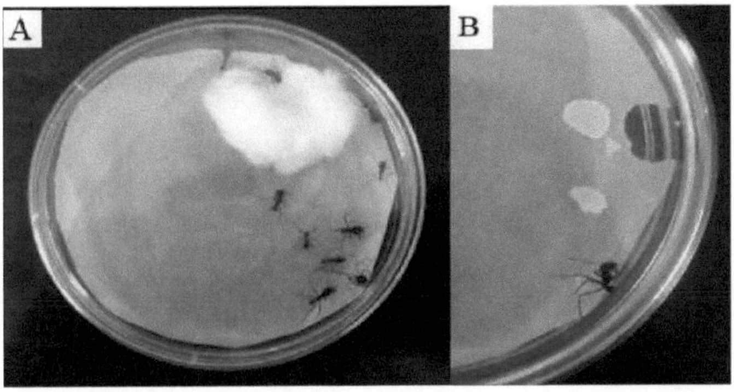

Source: Author (2017).

The bioassay showed that the isolates of Aspergillus sp., B. bassiana, B. brongniartii, M. anisopliae var. anisopliae and Paecilomyces sp. TP08 were pathogenic to leaf-cutting ants at the two concentrations of conidia used (1.0 x 10^4 conidia.mL^{-1} and 1.0 x 108 conidia.mL^{-1}).

The mortality of leaf-cutting ants by the fungi was confirmed by checking the extrusion of the pathogens (LOUREIRO; MONTEIRO, 2005) (**Figure 07**).

In the controls, there was no fungal growth on the cadavers of leaf-cutting ants, indicating that mortality in these groups was not the result of infection by fungi, but was possibly due to the stress caused by social isolation or even due to the senescence of the ants (RIBEIRO et al., 2012; PALMA, 2016; SILVA, 2017).

Figure 07. Extrusion of entomopathogenic fungi into leaf-cutter ant cadavers. **(A)** Aspergillus sp. **(B)** Beauveria bassiana. **(C)** Beauveria brongniartii. **(D)** Metarhizium anisopliae var. anisopliae. **(E)** Paecilomyces sp.

Source: Author (2017).

At a concentration of 1.0 x 10^{4} conidia.mL^{-1} , all the isolates, with the exception of Paecilomyces sp., showed mortality rates above 86 %. Only the Aspergillus sp. isolate showed a mortality rate of 100% at this concentration (**Figure 08**).

At a concentration of 1.0 x 10^{8} conidia.mL^{-1} , all the isolates, with the exception of Paecilomyces sp., showed mortality rates above 96 %. Three isolates (Aspergillus sp., B. brongniartii and M. anisopliae var. anisopliae) showed a mortality rate equal to 100% for this concentration (**Figure 08**).

Figure 08. Mortality (%) of leaf-cutting ants subjected to the control treatment (autoclaved distilled water + Tween 80 0.01 %) and to concentrations of 1.0 x 10^{4} conidia.mL^{-1} + Tween 80 0.01 % and 1.0 x 10^{8} conidia.mL^{-1} + Tween 80 0.01 % of different isolates of entomopathogenic fungi. Incubation period - 10 days, T = 25 ± 2 °C and relative humidity 70 ± 10 % (n=30).

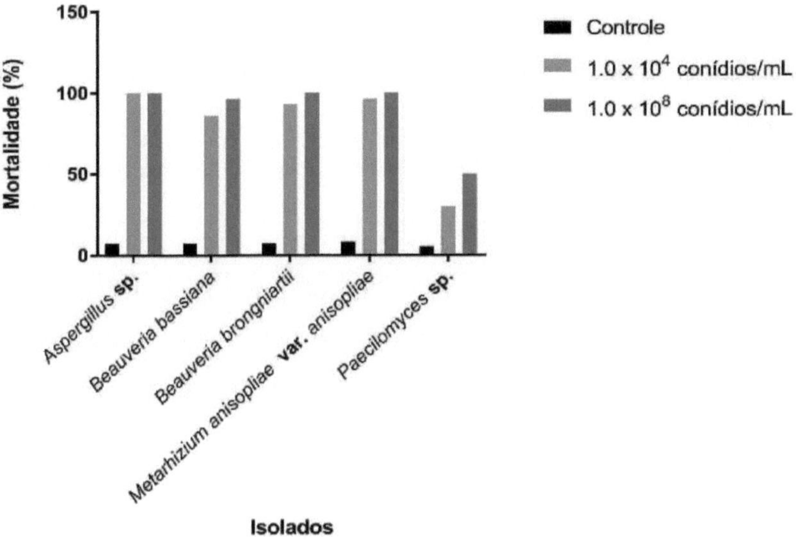

Source: Author (2017).

The data was statistically analyzed and the average mortality of leaf-cutting ants and the lethal time 50 (TL50) for each fungal isolate were obtained (**Tables 02** and **03**).

The results shown in **Table 02**, referring to the concentration of 1.0 x 104 conidia.mL^{-1} , suggest that there was a statistically significant difference between the isolates of Aspergillus sp. and Paecilomyces sp. TP08. At this concentration, the Aspergillus sp. isolate had the highest average mortality (9.99 ± 0.10) and the lowest TL50 (6.47 days). In contrast, the Paecilomyces sp. isolate had the lowest average mortality (4.80 ± 2.10) and the highest TL50 (10.01 days).

Table 02. Average mortality (± EP) and average lethal time (TL50) for the bioassay carried out with leaf-cutting ants subjected to a concentration of 1.0 x 104 conidia.mL^{-1} from five isolates of entomopathogenic fungi. Incubation period - 10 days (T = 25 ± 2 °C and relative humidity 70 ± 10 %) (n=30).

Isolated	**Average ant mortality (n=30) 1.0 x104 conidia.mL^{-1}**	**TL50 (Days)**
Metarhizium anisopliae	9.02 ± 0.42 ab*	7,33
Beauveria bassiana	8.89 ± 0.10 ab	9,01

Beauveria brongniartii	9.00 ± 0.32 ab	8,34
Paecilomyces sp. TP08	4,80 ± 2,10 b	10,01
Aspergillus sp.	9,99 ± 0,10 a	6,47
CV**	17,3	

*Means followed by the same letter do not differ by Tukey's test at 5% probability. Data not transformed. **CV= Coefficient of variation (%).

The results shown in **Table 03**, referring to the concentration of 1.0 x 10^8 conidia.mL^{-1} , suggest that there was a statistically significant difference between the isolates of Aspergillus sp. and Paecilomyces sp. TP08. *As* observed at a concentration of 1.0 x 104 conidia.mL^{-1} , the *Aspergillus* sp. isolate also had the highest average mortality (9.99 ± 0.00) and the lowest TL50 (6.01 days) at this concentration. Similarly, the *Paecilomyces* sp. isolate had the lowest average mortality (6.20 ± 1.30) and the highest TL50 (10.22 days).

Table 03. Average mortality (± EP) and average lethal time (TL50) for the bioassay carried out with leaf-cutting ants subjected to a concentration of 1.0 x 108 conidia.mL^{-1} from five isolates of entomopathogenic fungi. Incubation period - 10 days (T = 25 ± 2 °C and relative humidity 70 ± 10 %) (n=30).

Isolated	**Average ant mortality (n=30)** **1.0 x10^8 conidia.mL^{-1}**	**TL50 (Days)**
Metarhizium anisopliae	9.09 ± 0.20 ab*	6,93
Beauveria bassiana	8.99 ± 0.09 ab	8,67
Beauveria brongniartii	9.06 ± 0.02 ab	7,89
Paecilomyces sp. TP08	6,20 ± 1,30 b	10,22
Aspergillus sp.	9,99 ± 0,00 a	6,01
CV**	12,4	

*Means followed by the same letter do not differ by Tukey's test at 5% probability. Data not transformed. **CV = Coefficient of variation (%).

The variation in the virulence of the isolates observed in the bioassay with leaf-cutting ants can be explained mainly by the natural occurrence of inter- and intraspecific genetic variability in entomopathogenic fungi (ALVES, 1998; DIEHL-FLEIG et al., 1988 apud LOUREIRO; MONTEIRO, 2005; OLIVEIRA et al., 2004; CONCESCHI, 2017). This is reflected in differences in the production of enzymes and toxins, adhesion, speed of germination of conidia, cuticle penetration activity, as well as the fungi's ability to colonize and specificity (PACOLA-MEIRELLES; AZEVEDO, 1990; KLEESPIES; ZIMMERMANN, 1998 apud LOUREIRO et al., 2005a).

These differences in the characteristics and adaptations that each isolate possesses reinforce the importance of carrying out selection tests (GAUGLER et al., 1997 apud ALVES et al., 2009).

During the selection of an isolate, various characteristics of the pathogen are evaluated, such as host range, resistance to adverse environmental conditions, germination speed, conidia production, reproduction capacity, mycelial growth, pathogenicity and virulence. From a commercial point of view, ease of production and application, specificity, absence of toxicity, among others (MOINO JR.; ALVES, 1997; ALVES, 1998; RAMOS et al., 2004; LOUREIRO et al., 2005b; SANTORO et al., 2007; LA et al., 2013).

Knowing the TL_{50} of an isolate is important. From this data it is possible to draw some conclusions about the potency of pathogens, as well as the susceptibility of insects to the disease and even make predictions about pest control (OLIVEIRA, 2013).

Similarly to Silva (2001), the values for mortality and TL50 in this study are negatively correlated, i.e. as mortality increases, the time taken to kill half the population decreases (**Tables 02** and **03**).

According to Tamai et al. (2002a), when it comes to controlling agricultural pests, the speed with which a pathogen kills its host is considered a desirable characteristic. However, it is worth noting that the speed of action of the fungus depends on factors such as the dosage applied, the host species involved and the stage of development (XAVIER; ÂVILA, 2005; LOUREIRO et al., 2005a).

The ability to provide high final mortality is essential, since the main objective is to control as many individuals as possible (OLIVEIRA et al., 2002a; TAMAI et al., 2002a).

Determining the dosages or concentrations of a given pathogen to control insect pests is another important factor (MOINO JR.; ALVES, 1997).

Both concentrations of conidia used in this study (1.0 x 104 conidia.mL^{-1} and 1.0 x 108 conidia.mL^{-1}) were able to ensure the infection and progress of the disease in leaf-cutting ants. Furthermore, in some cases, increasing the concentration of conidia contributed to higher mortality in the insects.

Aspergillus sp. was the isolate that showed the highest average mortalities and the lowest TL50 for both concentrations of conidia tested in this study (**Tables 02** and **03**).

Some fungi belonging to the genus Aspergillus produce a variety of compounds that can be potent toxins against insects, making them potentially useful as mycopesticides (RIBEIRO et al., 2012).

Dornelas et al. (2016), in a bioassay with leaf-cutting ants Atta sexdens treated with a concentration of 1.0 x 10^7 conidia.mL^{-1} of an isolate of *Aspergillus niger* (T = 25 ± 2 °C; Relative humidity 70 ± 2.5%; 12 hours of photophase) (n = 20), found a TL50 of only 4.50 days, while in the present work, for an isolate of *Aspergillus* sp., obtained TL50 values of 6.47 and 6.01 days for concentrations of 1.0 x 104 conidia.mL^{-1} and 1.0 x 108 conidia.mL^{-1} , respectively. However, the mortality found by the authors for A. sexdens workers was 95% up to the tenth day of evaluation, whereas in the present study, 100% mortality was obtained for both conidial concentrations at the end of the ten days of evaluation.

In addition to the intrinsic qualities of the pathogens, there are some factors that must also be considered, such as the susceptibility, natural resistance, behavior, age and population density of the target pest (ALVES, 1998; PADULLA; ALVES, 2009; AGOSTINI et al., 2015; JUNIOR, 2017).

Insect groups differ in their susceptibility to fungal infections. These differences may be related to a variety of factors, including the presence or absence of symbionts, melanism linked to density or genetic variation (DUBOVSKIY et al., 2013).

Leaf-cutting ants of the genus *Atta* build large and complex nests, capable of holding millions of individuals (KLEINEIDAM et al., 2001). An *Atta vollenweideri* nest that is around 7.4 years old, for example, can have approximately 4 million individuals (JONKMAN, 1977 apud PIMENTA et al., 2007). The Density-Dependent Prophylaxis hypothesis relates the density of individuals to investment in defense. According to this hypothesis, organisms that live in high population densities tend to invest more in disease resistance mechanisms compared to those that live in low population densities, since the *per capita* risk of infection generally increases with increasing population (THOMPSON, et al., 2002; COTTER et al., 2008; WILSON; COTTER, 2008; GUIMARÃES, 2010).

Grooming behavior is common among social insects. Considered a defense mechanism, *grooming* is the act of cleaning oneself with one's paws using sweeping movements. This is a very useful strategy, which helps to eliminate infective units of pathogens and prevents the spread of diseases among members of the colony. It has been reported that ants, like termites, increase *grooming* when exposed to fungal conidia (MIYASHIRA, 2007; ZARZUELA, 2010; GANDRA, 2014; TRANTER; HUGHES, 2015).

The fungi must therefore be able to overcome or break through the social organization and defenses of leaf-cutting ants (MATTOSO, 2012).

When a particular pathogen comes into contact with the insect, one of the first barriers it encounters is the cuticle and, if ingested, the peritrophic membrane. When this first line of defense is overcome, the pathogen then has to deal with the host's humoral and cellular defenses (DUBOVSKIY et al., 2013; COUCEIRO, 2015).

Despite the activation of the host insect's immune system, evidenced by processes such as melanization, phagocytosis, nodulation and encapsulation, entomopathogenic fungi have managed to develop mechanisms that allow them to avoid such defenses, such as changes in the cell wall, production of immunomodulatory substances or toxins (COVA et al., 2009). *M. anisopliae*, for example, releases a protease capable of inhibiting the adhesion and phagocytosis activity of the plasmocytes of infected insects (SILVA, 2002).

The multiple mechanisms of action give entomopathogenic fungi a high capacity to prevent the host from developing resistance (DELGADO; MURCIA-ORDONEZ, 2011).

Trials such as the ones carried out in this study are necessary in order to understand the potential of the different genetic resources, contributing to the selection of the most appropriate isolates, which have satisfactory qualities and are suitable for the existing conditions, thus enabling their use in pest control programs (ALVES, 1998; ANDALÓ et al., 2004; PINTO, 2016).

5.4. Evaluation of the effect of the insecticide Fipronil on the vegetative growth and

sporulation (conidiogenesis) of different isolates of entomopathogenic fungi and classification *in terms of in vitro* toxicity

From the in vitro test, it was observed that the isolates of Aspergillus sp., B. bassiana, B. brongniartii, M. anisopliae var. anisopliae and Paecilomyces sp. TP08 treated with the insecticide Fipronil, at a concentration of 0.8 g/L, showed a reduction in vegetative growth compared to their respective controls (**Table 04**).

Aspergillus sp. and Paecilomyces sp. were the isolates that showed the lowest reduction in vegetative growth when exposed to the insecticide Fipronil, 21.61 ± 1.09 and 21.42 ± 2.08, respectively. The isolate of B. bassiana, on the other hand, showed the greatest reduction in vegetative growth of all the isolates of entomopathogenic fungi tested, around 80.90 ± 0.98 (**Table 04**).

This value was higher than that found by Rashid et al. (2012), who for this same concentration of Fipronil (0.8 g/L), obtained a reduction in vegetative growth (mm) of only 43.50 ± 2.60 % for the B. bassiana fungus (T= 25 ± 10 °C; Photophase 12 hours; Agar-Sabouraud-Dextrose medium) (n=9).

Table 04. Effect of the insecticide Fipronil on the vegetative growth (mm) of different isolates of entomopathogenic fungi. Incubation period - 10 days (Agar-Sabouraud-Dextrose medium; [Fipronil] = 0.8 g/L; T = 25 ± 2 °C; relative humidity 70 ± 10 %) (n=3) based on mean percentage ± SD.

Isolated	**Vegetative growth (mm) -CONTROL**	**Reduction in vegetative growth (%) CONTROL**	**Vegetative growth (mm) - FIPRONIL**	**Reduction in vegetative growth (%) FIPRONIL**
Metarhizium anisopliae	60.31 ± 0.49 bc*	39.69 ± 4.32 ab	32.09 ± 2.91 ab	68.00 ± 4.82 ab
Beauveria bassiana	39,80 ± 5,91 a	60,20 ± 3,70 c	19,10 ± 1,18 a	80,90 ± 0,98 c
Beauveria brongniartii	73,09 ± 1,03 b	26,91 ± 1,98 a	44,48 ± 4,01 b	55.52 ± 0.0 ab
***Paecilomyces* sp. TP08**	79,13 ± 1,55 b	20,87 ± 2,85 a	78,58 ± 3,55 c	21,42 ± 2,08 a

Aspergillus sp.	79,91 ± 5,98 b	20,09 ± 4,31 a	78,39 ± 2,81 c	21,61 ± 1,09 a

*Means followed by the same letter do not differ by Duncan's multiple range test ($P < 0.05$).

With regard to the second parameter evaluated, the in vitro test revealed that the isolates of Aspergillus sp., M. anisopliae var. anisopliae and Paecilomyces sp. TP08 treated with the insecticide Fipronil, at a concentration of 0.8 g/L, did not show a reduction in conidia production when compared to their respective controls (**Table 05**).

The isolates of B. bassiana and B. brongniartii treated with Fipronil, on the other hand, showed a decrease in the amount of conidia produced compared to their respective controls. However, there was no statistically significant difference between these treatments and their respective controls.

Table 05. Effect of the insecticide Fipronil on the sporulation (conidiogenesis) of different isolates of entomopathogenic fungi. Incubation period - 10 days (Agar-Sabouraud-Dextrose medium; [Fipronil] = 0.8 g/L; T = 25 ± 2 °C; relative humidity 70 ± 10 %) (n=3).

Isolated	**Sporulation** (x107conidia/mL) **CONTROL**	**-Sporulation (x 10^7 conidia/mL)** - FIPRONIL
Metarhizium anisopliae	4,50 ± 0,18 a	4,80 ± 1,19 a
Beauveria bassiana	5,10 ± 2,11 a	4,10 ± 1,81 a
Beauveria brongniartii	5,10 ± 1,93 a	4,50 ± 3,02 a
***Paecilomyces* sp. TP08**	9.60 ± 2.05 ab	17,00 ± 1,35 b
***Aspergillus* sp.**	19,00 ± 7,89 c	26.00 ± 3.18 bc

*Means followed by the same letter do not differ by Duncan's multiple range test ($P < 0.05$).

In view of the results obtained in relation to vegetative growth and conidiogenesis, the insecticide Fipronil was classified in terms of its in vitro toxicity to each of the entomopathogenic fungal isolates (**Table 06**).

Table 06. Classification of the insecticide Fipronil with regard to *in vitro* toxicity on *Aspergillus* sp., *Beauveria bassiana*, *Beauveria brongniartii*, *Metarhizium anisopliae* var. *anisopliae* and *Paecilomyces* sp. TP08. Incubation period - 10 days (Agar Sabouraud-Dextrose medium; [Fipronil] = 0.8 g/L; T = 25 ± 2 °C; relative humidity 70 ± 10 %) (n=3).

Isolated	**Values "T"**	**Classification***
Metarhizium anisopliae	95	Compatible
Beauveria bassiana	73	Compatible
Beauveria brongniartii	82	Compatible
Paecilomyces sp. TP08	161	Compatible
Aspergillus sp.	129	Compatible

*Classification in accordance with ALVES et al. (1998).

Chemical control and biological control have been used in association in integrated pest management (SILVA; NEVES, 2005). This strategy aims to increase efficiency and accelerate insect mortality by combining entomopathogenic fungi with sublethal doses of chemical insecticides, which can be advantageous in some cases where the isolated action of the microorganism is not entirely satisfactory (BENZ, 1971 apud WENZEL, 2005; ASI et al., 2010).

The joint application of insecticides and fungi helps to delay the expression of resistance in insect pests, leading to a significant reduction in the use of insecticides and reducing the exposure of non-target insects to them (FARGUES, 1975 apud AMBETHGAR, 2009). However, the use of entomopathogenic microorganisms and pesticides generally requires knowledge of the action of these products on microbial agents (BOTELHO; MONTEIRO, 2011).

Some insecticides can act as synergists in the development of insect diseases caused by fungi (PAVONE; DORTA, 2010). Others, however, can act to inhibit the vegetative growth, sporulation and conidiogenesis of microbial agents. In addition, they can cause mutations that can lead to a reduction in the virulence of fungi for a given pest (ALVES, 1998; TANZINI et al., 2002). It is therefore necessary to use compatible pesticides so that IPM is not compromised (ALMEIDA et al., 2003).

The toxicity of pesticides on fungal entomopathogens varies depending on the species or fungal lineage, the chemical nature of the active ingredient, the mode of action, the product formulation, among others, and different responses have been observed, ranging from synergistic, antagonistic to neutral (ALVES; LECUONA, 1998 apud SILVA et al., 2013b; ASI et al., 2010).

In the methodology proposed by Alves et al. (1998), the chemicals are added in previously established concentrations to the molten culture medium, i.e. not yet solidified, and the formula for classifying them *in terms of their in vitro* toxicity to entomopathogenic fungi is based on calculating the T factor (OLIVEIRA et al., 2003; SILVA et al., 2005).

In this study, the five entomopathogenic fungal isolates treated with the insecticide Fipronil at a concentration of 0.8 g/L showed a reduction in vegetative growth when compared to the control.

A possible explanation for these results was raised by Oliveira et al. (2002b). The authors suggest that the metabolization of the medium containing the formulation by the fungi can generate toxic residues which, when they accumulate, can block the pathways of important compounds for fungal growth.

According to Zimmerman (1975), however, inhibition of the vegetative growth of a fungus by a chemical product is not necessarily an indication of a reduction in its sporulation or conidial viability (CAZORLA; MORENO, 2010). Thus, if a fungal colony grows little, but produces many conidia, the rate of spread of the disease tends to be higher than that of a well-developed colony, but with less conidiogenesis (MOINO JR.; ALVES, 1998).

As mentioned by Moino Jr. & Alves (1998), conidiogenesis compared to vegetative growth is considered the most important parameter, since conidia are the fungal propagules that will act in the development of the disease in insect populations.

In this study, the insecticide Fipronil, at the concentration tested (0.8 g/L), did not compromise the conidiogenesis of the isolates of Aspergillus sp., M. anisopliae var. anisopliae and Paecilomyces sp. TP08, which had a higher quantity of conidia produced than the respective controls. And, despite the decrease observed in the amount of conidia of the isolates of B. *bassiana* and B. *brongniartii* treated with Fipronil, when compared to the controls, these values were not statistically significant (**Table 05**).

A possible explanation for the apparently conflicting results obtained *in* the *in vitro* test with Fipronil for the conidiogenesis parameter was raised by Moino Jr. & Alves (1998). One of the hypotheses suggested by the authors is that the fungus, faced with a toxic principle that alters its environment, makes a high reproductive effort, resulting in an increase in the production of conidia.

In view of the results regarding the vegetative growth and conidiogenesis of the five entomopathogenic fungal isolates and based on the formula proposed by Alves et al. (1998), Fipronil was classified as "compatible" *in the in vitro* compatibility test.

As mentioned by Tamai et al. (2002b), the high values of the "*T*" factor (> 60 = compatible) were determined by the good performance in the production of conidia, considering that the formula proposed by Alves et al. (1998) assigns 80% participation to this parameter in the final composition of the value.

Several studies have been carried out in order to understand the action of chemical products on entomopathogenic fungi (GUARANÀ, 2007). In general, toxicity tests for microorganisms are conducted in the laboratory (DEGRANDE et al., 2002 apud JÙNIOR, 2006).

The methodology proposed by Alves et al. (1998) is one of the most widely used in vitro methods and has been adopted in many experiments, especially in Brazil. *In vitro* studies have the advantage of exposing microbial agents as much as possible to the action of chemical products (SILVA et al., 2005; KAGIMURA et al., 2011). Some authors only evaluate vegetative growth and sporulation in in *vitro* tests, while others discuss and emphasize the importance of considering additional parameters, such as conidial germination (SILVA et al., 2005). Although vegetative growth and sporulation are important for colonizing the host and transmitting the fungus, according to some authors, germination should also be taken into account, as entomopathogenic fungi infect insects through the germination of conidia, either by ingestion or contact. If the germination of conidia is inhibited, it can be inferred that the affected fungus will not penetrate the host and therefore the disease will not develop (FAION, 2004; MERTZ et al., 2010).

Compatibility studies between phytosanitary products and entomopathogenic fungi are indispensable for integrated pest management. The findings of such studies allow farmers to select appropriate compounds, contributing to the preservation of pathogens and so that the deleterious effects of incompatible chemicals can be minimized (SILVA; NEVES, 2005; AMUTHA et al., 2010).

In an associated control strategy, priority should be given to using less harmful products, which are therefore more selective (LOUREIRO et al., 2002). When the product is compatible under laboratory conditions, there is strong evidence of its selectivity under field conditions. However, it is important to carry out experiments in semi-field and field conditions in order to assess the effectiveness of entomopathogenic fungi when associated with chemical insecticides for controlling insect pests (ALVES, 1998; RASHID et al., 2012).

6. Conclusions

S The isolates of *Aspergillus* sp., B. *bassiana, B. brongniartii, M. anisopliae* var. *anisopliae* and *Paecilomyces* sp. TP08 were pathogenic to leaf-cutting ants at concentrations of 1.0×10^{4} conidia.mL^{-1} and 1.0×10^{8} conidia.mL^{-1} .

S In some cases, the increased concentration of fungal conidia contributed to higher insect mortality.

S Aspergillus sp. was the isolate that showed the highest average mortalities and the lowest TL_{50} for both concentrations of conidia tested.

S The five entomopathogenic fungal isolates treated with the insecticide Fipronil at a concentration of 0.8 g/L showed a reduction in vegetative growth (mm).

S The insecticide Fipronil, at the concentration tested, did not significantly compromise the conidiogenesis of the five entomopathogenic fungal isolates.

S The insecticide Fipronil was classified as "compatible" *in terms* of *in vitro* toxicity against the five entomopathogenic fungal isolates.

Finally, the bioassay described in this study allowed the discovery of isolates with great potential to be evaluated in the field for the biological control of leaf-cutting ants, respecting, of course, the specific aspects of the target ecosystem, the fungal isolate and the leaf-cutting ants.

7. REFERENCES

ABARCA, M. L. Taxonomy and identification of species implicated in nosocomial aspergillosis. **Revista Iberoamericana de Micologia**, v. 17, p. 79-84, 2000.

ABREU, J. M. de; DELABIE, J. H. C. Control of leaf-cutting ants in cocoa plantations. **Revista Theobroma**, v. 16, n. 4, p. 199-211, 1986.

AGALI, A. E. A. E.; MOHAMMED, E. A. E. R.; HASSAN, A. E. W. Effect of entomopathogenic fungi Metarhizium anisopliae var acridum and Beauveria bassiana on survival of Anopheles arabiensis Patton and Culex quinquefasciatus Say mosquito larvae. **Sudan Journal of Science**, v. 9, n. 2, p. 26-41, 2017.

AGOSTINI, T. T.; AGOSTINI, L. T.; DUARTE, R. T.; VOLPE, H. X. L.; SALAS, C.; POLANCZYK, R. A. Efficiency of entomopathogenic fungi for the control of *Sitophilus oryzae* L. (Coleoptera: Curculionidae) under laboratory conditions. **Comunicata Scientiae**, v. 6, n. 1, p. 90-96, 2015.

ALBUQUERQUE, A. C.; PEREIRA, K. C. A.; CUNHA, F. M.; VEIGA, A. F. S. L.; ATHAYDE, A. C. R.; LIMA, E. A. L. A. Pathogenicity of *Metarhizium anisopliae* var. *anisopliae* and *Metarhizium anisopliae* var. *acridum* on *Nasutitermes coxipoensis* (Holmgren) (Isoptera: Termitidae). **Neotropical Entomology**, v. 34, n. 4, p. 585-591, 2005.

ALEXOPOULOS, C.J.; MIMS, C.W. **Introductory mycology**. 3 ed. New York: John Wiley & Sons Inc., 1979.

ALMEIDA, J. E. M.; FILHO, A. B.; LAMAS, C.; LEITE, L. G.; TRAMA, M.; SANO, A. H. Evaluation of the compatibility of pesticides in the conservation of entomopathogenic microorganisms in the management of coffee tree pests. **Arquivos do Instituto Biológico**, v. 70, n. 1, p.79-84, 2003.

ALMEIDA, J. T. S.; MEDICI, L. O.; AGUIAR-MENEZES, E. L. Efficiency and principle of operation of a conical physical barrier against quenquéns. **Floresta**, v. 43, n. 4, p. 633-642, 2013.

ALMEIDA, R. N. A.; PENAFLOR, M. F. G. V.; SIMOTE, S. Y.; BUENO, O. C.; HEBLING, M. J. A.; PAGNOCCA, F. C.; FERNANDES, J. B.; VIEIRA, P. C.; SILVA, M. F. G. F. Toxicity of Substances Isolated from *Helietta puberula* RE Fr. (Rutaceae) to the Leaf-cutting Ant *Atta sexdens* L. (Hymenoptera: Formicidae) and the Symbiotic Fungus *Leucoagaricus gongylophorus* (Singer) Moller. **BioAssay**, v. 2, n. 2, p. 1-8, 2007.

ALVES-FAVA, J.; NAVAS, C.; RIBEIRO, P. L. Production and didactic application of the video "as saùvas: uma sociedade de ants". **Revista Cultura e Extensao USP**, v. 5, p. 45-53, 2011.

ALVES, G. F. **Solubilization of rock phosphate by** Aspergillus niger. 2012. Dissertation (Master's Degree in Chemical Engineering) - Federal University of Uberlândia, Uberlândia, Minas Gerais.

ALVES, R. T.; FARIA, M. **Pequeno Manual sobre fungos entomopatogênicos**. 1. ed. Planaltina, Distrito Federal: Embrapa Cerrados, 2010.

ALVES, S. B. **Microbial control of insects**. 2. ed. Piracicaba: FEALQ, 1998.

ALVES, S.B.; LECUONA, R.E. **Epizootiology applied to microbial control**. In: ALVES, S. B. Controle microbiano de insetos. 2. ed. Piracicaba: FEALQ, 1998, p. 97-157.

ALVES, S. B.; MOINO JR., A.; ALMEIDA, J. E. M. **Phytosanitary products and entomopathogens**. In: ALVES, S. B. Controle microbiano de insetos. 2. ed. Piracicaba: FEALQ, 1998, p. 217-238.

ALVES, V. S.; MOINO JR., A.; SANTA-CECILIA, L. V. C.; ANDALÓ, V.;

SOUZA, G. C. Pathogenicity of entomopathogenic nematodes to the coffee root mealybug Dysmicoccus texensis (Tinsley) (Hemiptera: Pseudococcidae) in the laboratory. **Arquivos do Instituto Biológico**, v. 76, n. 1, p. 67-73, 2009.

ALVIRA, C. O. A. **Dinamica y diversidad de la hormiga cortadora (Genero** Atta**) en**

different uses of the soil, in the El Cabuyal footpath in the Municipality of De la Plata, Department of Huila. 2014. Monograph (Forestry Engineering) - Universidad Nacional Abierta y a Distancia, Colombia.

ALVIRA, C. O. A.; ACOSTA, D. R. Diversidad de hormigas cortadoras de hojas en tres usos de suelo en la vereda El Cabuyal del Municipio de La Plata, Huila. **Revista Nova**, v. 1, n. 1, p. 38-47, 2015.

AMBETHGAR, V. Potential of entomopathogenic fungi in insecticide resistance management (IRM): A review. **Journal of Biopesticides**, v. 2, n. 2, p. 177-193, 2009.

AMUTHA, M.; BANU, J. G.; SURULIVELU, T.; GOPALAKRISHNAN, N. Effect of commonly used insecticides on the growth of white muscardine fungus, Beauveria bassiana under laboratory conditions. **Journal of Biopesticides**, v. 3, n. 1, p. 143 - 146, 2010.

ANDALÓ, V.; MOINO JR., A.; SANTA-CECILIA, L. V. C.; SOUZA, G. C. Selection of fungal isolates and entomopathogenic nematodes for the coffee root mealybug Dysmicoccus texensis (Tinsley). **Arquivos do Instituto Biológico**, v. 71, n. 2, p. 181-187, 2004.

ARAGÓN-GARCIA, A.; PÉREZ-TORRES, B. C.; CALDERÓN-GARCÌA, G.; CASTILLO-HERNÂNDEZ, D.; ARAGÓN-SÀNCHEZ, M.; RAMÓN, D. J. Uso del complejo de hongos de la tortilla de maiz (Zea *mays* L.) para el combate de la hormiga arriera Atta mexicana (Smith 1858) (Hymenoptera: Formicidae). **Entomologia mexicana**, v. 3, p. 153-158, 2016.

ARAÙJO, M. S.; DELLA LUCIA, T. M. C.; RIBEIRO, G. A.; KASUYA, M. C. M. Impact of Controlled Burning of Sugarcane on Nesting and Establishment of Colonies of Atta bisphaerica Forel (Hymenoptera: Formicidae). **Neotropical Entomology**, v. 32, n. 4, p. 685-691, 2003.

ARAÙJO, M. S.; RODRIGUES, C. A.; OLIVEIRA, M. A.; JESUS, F. G. Biological control of leaf-cutting ants: the case of predation of Atta spp. females by Canthon virens. **Revista de Agricultura Neotropical**, v. 2, n. 3, p. 8-12, 2015.

ARISMENDY, E. L.; PERALTA, S. O. Metarhizium anisopliae and Trichoderma viride control colonies of Atta cephalotes in the field better than a chemical insecticide. **Revista Colombiana de Biotecnologia**, v. 4, n. 1, p. 71-78, 2002.

ASI, M. R.; BASHIR, M. H.; AFZAL, M.; ASHFAQ, M.; SAHI, S. T. Compatibility of entomopathogenic fungi, Metarhizium anisopliae and Paecilomyces fumosoroseus with selective insecticides. **Pakistan Journal of Botany**, v. 42, n. 6, p. 4207-4214, 2010.

AUBAD, P. **Plantas usadas por las comunidades indigenas Ticuna del PNN Amacayacu para el control de la hormiga cortadora: Evaluación biológica de metabolitos secundarios**. 2010. Master's Degree - National University of Colombia.

AYLWARD, F. O.; BURNUM-JOHNSON, K. E.; TRINGE, S. G.; TEILING, C.; TREMMEL, D. M.; MOELLER, J. A.; SCOTT, J. J.; BARRY, K. W.; PIEHOWSKI, P. D.;

NICORA, C. D.; MALFATTI, S. A.; MONROE, M. E.; PURVINE, S. O.; GOODWIN, L. A.; SMITH, R. D.; WEINSTOCK, G. M.; GERARDO, N. M.; SUEN, G.; LIPTON, M. S.; CURRIE, C. R. Leucoagaricus gongylophorus Produces Diverse Enzymes for the Degradation of Recalcitrant Plant Polymers in Leaf-Cutter

Ant Fungus Gardens. **Applied and Environmental Microbiology**, v. 79, n. 12, p. 3770-3778, 2013.

BADII, M. H.; ABREU, J. L. Control biológico una forma sustentable de control de plagas. **Daena: International Journal of Good Conscience**, v. 1, n. 1, p. 82-89, 2006.

BALE, J. S.; LENTEREN, J. C.; BIGLER, F. Biological control and sustainable food production. **Philosophical Transactions of the Royal Society B**, v. 363, p. 761-776, 2008.

BARBOSA, R. H.; KASSAB, S. O.; PEREIRA, F. F.; ROSSONI, C.

Chemical and biological control of *Mahanarva fimbriolata Stài,* 1854 (Hemiptera: Cercopidae) for sugarcane producing regions of Mato Grosso do Sul.

Ambiência, v. 11, n. 1, p. 247-255, 2015.

BARBOZA, M. R.; SILVA, D. N.; LUSTOSA, S. B. C.; HIROSE, E. Pathogenicity of the entomopathogenic fungus *Beauveria bassiana* on the bed bug *Collaria scenica* (Hemiptera: Miridae). **Ambiência**, v. 7, n. 3, p. 473-480, 2011.

BASTOS, C. S.; TORRES, J. B. **Biological Control as an Option in Cotton Pest Management**. 1. ed. Campina Grande, Paraiba: Embrapa, 2003.

BAUTISTA-GALVEZ, A.; GONZALEZ-CORTES, N. Tres dosis de *Metarhizium anisopliae* sobre la mosca pinta *(Aeneolamia spp.)* em cana de azùcar en la región de los rios, estado de tabasco. **Universidad y Ciencia**, v. 21, n. 41, p. 37-40, 2005.

BENZ, G. **Synergism of micro-organisms and chemical insecticides**. In: BURGES, H. D.; HUSSEY, N. W. Microbial control of insects and mites. New York: Academic Press, 1971, p. 327-355.

BERGAMIN FILHO, A.; KIMATI, H.; AMORIM, L. **Manual de Fitopatologia**. 3. ed. v. 1. Sao Paulo: Agronòmica Ceres, 1995.

BHAN, S.; SHRANKHLA; MOHAN, L.; SRIVASTAVA, C. N. Larvicidal toxicity of Temephos and entomopathogenic fungus, *Aspergillus flavus* and their synergistic activity against malaria vector, *Anopheles stephensi*. **Journal of Entomology and Zoology Studies**, v. 1, n. 6, p. 55-60, 2013.

BOARETTO, M, A. C.; FORTI, L. C. Perspectives on the control of leaf-cutting ants. **IPEF Technical Series**, v. 11, n. 30, p. 31-46, 1997.

BONALDO, S. M.; PASCHOLATI, S. F. Effect of partially purified fractions of *Saccharomvces cerevisiae* on conidia germination and appressoria formation by Colletotrichum sublineolum and Colletotrichum lagenarium. **Summa Phytopathologica**, v. 33, n. 3, p. 233-238, 2007.

BORGES, L. R. **Analysis of microbiological quality (molds and yeasts) in yerba mate (**Ilex ***paraguariensis*** **St. Hil.) and identification of potentially mycotoxigenic fungi**. 1999. Monograph (Biological Sciences) - Federal University of Paranà, Curitiba.

BORGES, L. R.; NOVA, M. X. V. Association of chemical insecticides and entomopathogenic fungi in Integrated Pest Management - a review. **Ambiência**, v. 7, n. 1, p. 179-190, 2011.

BORJA, G. E. M. Eradication or integrated management of the neotropical myiases of the Americas? **Pesq. Vet. Bras.**, v. 23, n. 32, p. 131-138, 2003.

BOT, A. N. M.; ORTIUS-LECHNER, D.; FINSTER, K.; MAILE, R.; BOOMSMA, J. J. Variable sensitivity of fungi and bacteria to compounds produced by the metapleural glands of leaf-cutting ants. **Insectes Sociaux**, v. 49, p. 363-370, 2002.

BOTELHO, A. A. A.; MONTEIRO, A. C. Sensitivity of entomopathogenic fungi to agrochemicals used in sugarcane management. **Bragantia**, v. 70, n. 2, p. 361369, 2011.

BOTERO, C. E. G.; MARiN, P. M.; MACHADO, P. B. Claves para el éxito del hongo Beauveria bassiana como controlador biológico de la broca del café. **Avances Técnicos Cenicafé**, v. 384, p. 1-8, 2009.

BRAZIL. Ministry of Agriculture, Livestock and Supply (2010). System of phytosanitary pesticides. Retrieved from <http://extranet.agricultura.gov.br/agrofit_cons/principal_agrofit_cons>.

BRECHELT, A. **El Manejo Ecológico de Plagas y Enfermedades**. 1. ed. Santiago de Chile, Chile: Red de Acción en Plaguicidas y sus Alternativas para América Latina (RAP-AL), 2004.

BRITTO, J. S.; FORTI, L. C.; OLIVEIRA, M A.; ZANETTI, R.; WILCKEN, C. F.; ZANUNCIO, J. C.; LOECK, A. E.; CALDATO, N.; NAGAMOTO, N. S.; LEMES, P. G.; CAMARGO, R. S. Use of alternatives to PFOS, its salts and PFOSF for the control of leaf-cutting ants Atta and Acromyrmex. **International Journal of Research in Environmental Studies**, v. 3, p. 11-92, 2016.

BUENO, F. C. **Selection of active ingredients for the development of toxic baits for the control of leaf-cutting ants (Hymenoptera: Formicidae)**.

2013. Thesis (PhD in Biological Sciences) - Institute of Biosciences, Paulista State University, Rio Claro.

CANEDO, V.; AMES, T. **Manual de laboratorio para el manejo de hongos entomopatógenos**. Lima, Peru: Centro Internacional de la Papa (CIP), 2004.

CAFARO, M. J.; POULSEN, M.; LITTLE, A. E. F.; PRICE, S. L.; GERARDO, N. M.; WONG, B.; STUART, A. E.; LARGET, B.; ABBOT, P.; CURRIE, C. R. Specificity in the symbiotic association between fungus-growing ants and protective Pseudonocardia bacteria. **Proceedings of the Royal Society B**, v. 278, p. 1814-1822, 2011.

CAMARGO, R. S.; FONSECA, J. A.; LOPES, J. F. S.; FORTI, L. C. Influence of the environment on the development of initial colonies of leaf-cutting ants (Atta sexdens rubropilosa). **Ciência Rural**, v. 43, n. 8, p. 1375-1380, 2013.

CANTARELLI, E. B.; COSTA, E. C.; OLIVEIRA, L. S.; PERRANDO, E. R. Effect of different doses of the formicide "citromax" on the control of Acromyrmex lundi (HYMENOPTERA: FORMICIDAE). **Ciência Florestal**, v. 15, n. 3, p. 249-253, 2005.

CANTARELLI, E. B.; COSTA, E. C.; PEZUTTI, R.; OLIVEIRA, L. S. Quantification of losses in the development of *Pinus taeda* after attack by leaf-cutting ants. **Ciência Florestal**, v. 18, n. 1, p. 39-45, 2008.

CANTÙ-RUIZ, A. L.; GALVÂN-QUINTERO, A. O.; MAR-SOLIS, L. M. Aplicaciones biotecnológicas en el control biológico. **Artrópodos y Salud**, v. 7, n. 1, p. 54-70, 2017.

CARRILLO-RAYAS, M. T.; BLANCO-LABRA, A. Potencial y Algunos de los Mecanismos de Acción de los Hongos Entomopatógenos para el Control de Insectos Plaga. **Acta Universitaria**, v. 19, n. 2, p. 40-49, 2009.

CARVALHO, L. I. C. Aspergillus **and Aspergillosis - challenges in combating the disease**.

2013. Dissertation (Integrated Master's Degree in Pharmaceutical Sciences) - Universidade Fernando Pessoa, Porto.

CARVALHO, N. L.; BARCELLOS, A. L. Adoption of Integrated Pest Management based on perception and environmental education. **Revista Eletrônica em Gestâo, Educaçâo e Tecnologia Ambiental**, v. 5, n. 5, p. 749-766, 2012.

CASSIANO, J. A.; DESTÉFANO, R. H. R.; BARACHO, M. S.; NAAS, I. A.; SALGADO, D. D. Adhesion analysis of the entomopathogenic fungus *Metarhizium anisopliae* for the control of *Alphitobius diaperinus* in poultry facilities. **Brazilian Journal of Veterinary Research and Animal Science**, v. 45, n. 5, p. 348-353, 2008.

CASTELLANOS-MOGUEL, J.; CRUZ-CAMARILLO, R.; ARANDA, E.; MIER, T.; TORIELLO, C. Relationship between protease and chitinase activity and the virulence of Paecilomyces fumosoroseus in Trialeurodes vaporariorum (Hemiptera: Aleyrodidae). **Revista Mexicana de Micologia**, v. 28, p. 71-80, 2008.

CASTILHO, A. M. C.; FRAGA, M. E.; AGUIAR-MENEZES, E. L.; ROSA, C. A.

R. Selection of isolates of *Metarhizium anisopliae* and *Beauveria bassiana* pathogenic to soldiers of *Atta bisphaerica* and *Atta sexdens rubropilosa* under laboratory conditions.

Ciência Rural, v. 40, n. 6, p. 1243-1249, 2010.

CASTILLO, C. E.; CANIZALEZ, L. M.; VALERA, R.; GODOY, J. C.; GUEDEZ, C.; OLIVAR, R.; MORILLO, S. Morphological characterization of *Beauveria bassiana*, isolated from different insects in Trujillo - Venezuela. **Revista Academia**, v. 11, n. 23, p. 275-281, 2012.

CASTILLO, S. Y.; VEGA, J. A. B.; LEZCANO, J.; PIEPENBRING, M.;

CACERES, O. Hongos entomopatógenos asociados a insectos recolectados en plantaciones de café en el oeste de Panamâ. **Tecnociencia**, v. 15, n. 2, p. 29-39, 2013.

CAZORLA, D.; MORENO, P. M. Compatibilidad de 13 aislamientos de *Beauveria bassiana*

pathogens for *Rhodnius prolixus* (Triatominae) with chemical insecticides. **Boletin de Malariologia y Salud Ambiental**, v. 50, n. 2, 2010.

CELIS, S. U. **Fóridos (Diptera: Phoridae) asociados al hâbitat de hormigas cortadoras de hojas (**Atta cephalotes y Acromyrmex octospinosus**) y sus patrones de localización en**

un bosque seco tropical Andino. 2013. Dissertation (Master's Degree in Science-Entomology) - Universidad Nacional de Colombia.

CHAGURI, J. L. **Effects of exposure to the pesticide fipronil on pressure changes in awake rats**. 2016. Dissertation (Master's Degree in Pharmacology and Biotechnology) - Institute of Biosciences, Universidade Estadual Paulista "Jûlio de Mesquita Filho", Botucatu.

CHAVEZ, B. Y.; ROJAS, J. C.; BARRERA, J. F.; GÓMEZ, J. Evaluación de la patogenicidad de *Beauveria bassiana* sobre *Pachycoris torridus* en Laboratorio.

Southwestern Entomologist, v. 41, n. 3, p. 783-790, 2016.

CONCESCHI, M. R. **Parameters to be considered in the spraying of the fungus** Isaria fumosorosea **for the management of** Diaphorina citri. 2017. Thesis (Doctorate in Sciences) - University of Sao Paulo, Piracicaba.

CORRÊA, M. M.; SILVA, P. S. D. WIRTH, R.; TABARELLI, M.; LEAL, I. R. How leaf-cutting ants impact forests: drastic nest effects on light environment and plant assemblages. **Oecologia**, v. 162, p. 103-115, 2010.

CRAVANZOLA, F.; PIATTI, P.; BRIDGE, P. D.; OZINO, I. O. Detection of genetic polymorphism by RAPD-PCR in strains of the entomopathogenic fungus Beauveria brongniartii isolated from the European cockchafer (Melolontha spp.). **Letters in Applied Microbiology**, v. 25, p. 289-294, 1997.

COSTA, F. D.; VERZELETTI, F. B.; WAGNER, R. Isolation and identification of aflatoxins B1 and B2 from Aspergillus parasiticus in food. **Cadernos da Escola de Saùde**, v. 11, p. 65-78, 2014.

COTTER. S. C.; MYATT, J. P.; BENSKIN, C. M. H.; WILSON, K. Selection for cuticular melanism reveals immune function and life-history trade-offs in Spodoptera littoralis. **Journal of Evolutionary Biology**, v. 21, p. 1744-1754, 2008.

COUCEIRO, J. C. **Action of entomopathogenic fungi on the mortality and immune system of leaf-cutting ants**. 2015. Dissertation (Postgraduate Diploma in Entomology) - Federal University of Viçosa, Viçosa, Minas Gerais.

COVA, L. J.; SCORZA, J. V.; GARCiA, D. E.; CANIZÂLEZ, L. M.; GUEDEZ, C. C.; MAFFEY, M.; MEDINA, M. G. In vitro pathogenicity of Beauveria brongniartii (Sacc.) Petch on Musca domestica (L.) as a possible biological control strategy in livestock areas.

Zootecnia Tropical, v. 27, n. 2, p. 113-120, 2009.

CRUZ, I. **Biological control of pests in the cultivation of corn for the production of canned corn (minimilho), by means of parasitoids and predators**. 1. ed. Sete Lagoas, Minas Gerais: Embrapa Milho e Sorgo, 2007.

CRUZ, J. C.; MAGALHAES, P. C.; FILHO, I. A. P.; MOREIRA, J. A. A. **Maize: the producer asks, Embrapa answers**. Brasilia, DF: Embrapa, 2011.

CRUZ, J. M.; NOGUEIRA, S. B.; PEREIRA, A. R.; MEWES, B. O. Adaptation of a motorcycle for thermonebulization in the control of leafcutter ants (Atta spp.) in Cerrado reforested areas. **Revista àrvore**, v. 8, n. 2, p. 104-111, 1984.

CURRIE, C. R.; BOT, A. N. M.; BOOMSMA, J. J. Experimental evidence of a tripartite mutualism: bacteria protect ant fungus gardens from specialized parasites. **OIKOS**, v. 101, p. 91-102, 2003.

DALZOTO, P. R.; UHRY, K. F. Biological control of pests in Brazil by means of Beauveria bassiana (Bals.) Vuill. **Biológico**, v. 71, n. 1, p. 37-41, 2009.

DAMIN, S.; VILANI, A.; FREITAS, D.; KRASBURG, C.; QUEIROZ, J. A.; KAGIMURA, F. Y.; ONOFRE, S. B. Action of fungicides on the growth of the entomopathogenic fungus *Metarhizium* sp. **Revista Acadêmica: Ciências Agrârias e Ambientais**, v. 9, n. 1, p. 41-49, 2011.

DÂNGELO, R. A. C. **Actinomycetes and symbiotic fungi: implications for leaf-cutting ants of the genus *Acromyrmex* Mayr, 1865**. 2011. Dissertation (Postgraduate Diploma in Entomology) - Federal University of Viçosa, Viçosa, Minas Gerais.

DANS, D.; ANGLADA, M. M.; MAIDANA, A. Characterization of leaf-cutting ant damage in *sorghum (Sorghum* spp) cultivation. **Revista Cientifica Agropecuâria**, v. 13, n. 1, p. 7-15, 2009.

DAUBER, J., WOLTERS, V. Microbial activity and functional diversity in the mounds of three different ant species. **Soil Biol. Biochem.**, v. 32, p. 93-99, 2000.

DEGRANDE, P. E.; REIS, P. R.; CARVALHO, G. A.; BELARMINO, L.C.; **Methodology for evaluating the impact of pesticides on natural enemies**. In: PARRA, J. R.; BOTELHO, P. S. M.; CÔRREA-FERREIRA, B. S.; BENTO, J. M. Controle Biològico no Brasil:

Parasitóides e predadores. Sao Paulo: Manole, 2002, p. 71-86.

DELGADO, P. A. M.; MURCIA-ORDONEZ, B. Hongos entomopatógenos como alternativa para el control biológico de plagas. **Ambi-Agua**, v. 6, n. 2, p. 77-90, 2011.

DELLA LUCIA, T. M. C. **Hormigas de importancia económica en la región Neotropical.** In: FERNANDEZ, F. Introducción a las hormigas de la región Neotropical. Bogotà, Colombia: Instituto de Investigación de Recursos Biológicos Alexander von Humboldt, 2003.

DELLA LUCIA, T. M. C.; GANDRA, L. C.; GUEDES, R. N. C. Managing leafcutting ants: peculiarities, trends and challenges. **Pest Management Science**, v. 70, p. 14-23, 2014.

DELLA LUCIA, T. M. C.; OLIVEIRA, M. A.; ARAÙJO, M. S.; VILELA, E. F. Evaluation of the non-preference of the leaf-cutting ant *Acromyrmex subterraneus subterraneus* Forel to the cutting of *Eucalyptus*. **Revista Arvore**, v. 19, n. 1, p. 92-99, 1995.

DiAZ, M. P.; MACiAS, A. F.; NAVARRO, S. R.; TORRE, M. Mecanismo de acción de los hongos entomopatógenos. **Interciencia**, v. 31, n. 12, p. 856-860, 2006.

DIEHL-FLEIG, E.; SILVA, M. E.; PACHECO, M. R. M. Pathogenicity tests of the entomopathogenic fungi Beauveria bassiana and Metarhizium anisopliae on Atta sexdens piriventris (Santschi, 1919) at different temperatures. **Ciência e Cultura**, v. 40, n. 11, p. 1103-1105, 1988.

DIEHL, E.; DIEHL-FLEIG, E.; ALBUQUERQUE, E. Z. Occurrence of Attini (Formicidae) in two geomorphological provinces of Rio Grande do Sul, Brazil, **Brazilian Journal of Agriculture**, v. 92, n. 1, p. 66-74, 2017.

DIODATO, M. A. **Natural occurrence, laboratory and field tests of** Beauveria bassiana **(Bals.) Vuill, on** Sirex noctilio **F., a pest of** Pinus taeda **L.** 1992. Dissertation (Post-graduation in Forest Engineering) - Federal University of Paranà, Curitiba.

DOLCI, P.; GUGLIELMO, F.; SECCHI, F.; OZINO, O. I. Persistence and efficacy of Beauveria brongniartii strains applied as biocontrol agents against Melolontha melolontha in the Valley of Aosta (northwest Italy). **Journal of Applied Microbiology**, v. 100, p. 1063-1072, 2006.

DORNELAS, A. S. P.; SARMENTO, R. A.; SANTOS, G. R.; NASCIMENTO, M. O.; SOUZA, D. J. Filamentous Fungi Associated with the Species Atta sexdens (Linnaeus) and Atta laevigata (F. Smith) (Hymenoptera: Formicidae). **EntomoBrasilis**, v. 9, n. 1, p. 26-30,

2016.

DUARTE, A. P. M.; ATTILI-ANGELIS, D.; BARON, N. C.; FORTI, L. C.; PAGNOCCA, F. C. Leaf-cutting ants: an unexpected microenvironment holding human opportunistic black fungi. **Antonie van Leeuwenhoek**, v. 106, p. 465-473, 2014.

DUBOVSKIY, I. M.; WHITTEN, M. M. A.; YAROSLAVTSEVA, O. N.; GREIG, C.;

KRYUKOV, V. Y.; GRIZANOVA, E. V.; MUKHERJEE, K.; VILCINSKAS, A.; GLUPOV, V. V.; BUTT, T. M. Can Insects Develop Resistance to Insect Pathogenic Fungi? **Plos One**, v. 8, n. 4, p. 1-9, 2013.

DUNCAN, D. B. Multiple range and multiple F tests. **Biometrics**, v. 11, p. 1-42, 1955.

ECKARD, S.; BACHER, S.; ENKERLI, J.; GRABENWEGER, G. A simple in vitro method to study interactions between soil insects, entomopathogenic fungi, and plant extracts. **Entomologia Experimentalis et Applicata**, v. 163, p. 315-327, 2017.

EHLER, L. E. Perspectives Integrated pest management (IPM): definition, historical development and implementation, and the other IPM. **Pest Management Science**, 2006.

ELENA, K.; PAPAS, A. C. Pathogenicity and vegetative compatibility of Fusarium oxysporum in alfalfa. **Journal of Phytopathology**, v. 150, p. 495-499, 2002.

ESTRADA, C.; WCISLO, W. T.; VAN BAEL, S. A. Symbiotic fungi alter plant chemistry that discourages leaf-cutting ants. **New Phytologist**, v. 198, p. 241-251, 2013.

ESTRADA, R. M. H.; SERMENO-CHICAS, J. M. Los Zompopos de mayo en El Salvador. **Bioma**, n. 7, p. 13-18, 2013.

EVANS, H. C. Uso actual y potencial de los hongos entomopatógenos para el control biológico de artrópodos plagas. **Revista Palmas**, v. 14, n. 1, p. 37-46, 1993.

FAIA, A. M. **Isolation and identification of filamentous fungi and yeasts in some points of a water distribution network**. 2011. Dissertation (Master's Degree in Cell Biology and Biotechnology) - University of Lisbon, Lisbon, Portugal.

FAION, M. **Toxicity of pesticides used to control** Bemisia tabaci **biotype B on entomopathogenic fungi**. 2004. Dissertation (Master of Science) - University of Sao Paulo, Piracicaba, Sao Paulo.

FAN, J.; XIE, Y.; XUE, J.; LIU, R. The effect of Beauveria brongniartii and its secondary metabolites on the detoxification enzymes of the pine caterpillar, Dendrolimus tabulaeformis. **Journal of Insect Science**, v. 13, n. 44, p. 1-13, 2013a.

FAN, J.; XIE, Y.; XUE, J.; ZHANG, Y.; YANG, Q. Cellular Apoptosis of Hemocytes from Dendrolimus tabulaeformis Tsai et Liu Larvae Induced with the Secondary Metabolites of Beauveria brongniartii (Sacc.) Petch. **PLOS ONE**, v. 8, n. 8, p. 1-10, 2013b.

FARDER-GOMES, C. F.; OLIVEIRA, M. A.; GONÇALVES, P. L.; GONTIJO, L.

M.; ZANUNCIO, J. C.; BRAGANÇA, M. A. L.; PIRES, E. M. Reproductive ecology of phorid parasitoids in relation to the head size of leaf-cutting ants Atta sexdens Forel.

Bulletin of Entomological Research, v. 107, p. 487-492, 2017.

FARGUES, J. Elude experimentale dans la nature de lutilisation combine de Beauveria bassiana et d'insecticides a dose reduite contre Leptinotarsa decemlineata. **Annales de Zoologie Ecologie Animale**, v. 7, p. 247-264, 1975.

FARJI-BRENER, A. G.; MEDINA, C. A. The Importance of Where to Dump the Refuse: Seed Banks and Fine Roots in Nests of the Leaf-Cutting Ants Atta cephalotes and A. colombica. **Biotropica**, v. 32, n. 1, p. 120-126, 2000.

FATU, A.; DINU, M.; CIORNEI, C. ANDREI, A. Biological control of Melolontha melolontha L. larvae with entomopathogenic bioinsecticide based on Beauveria brongniartii. **AgroLife Scientific Journal**, v. 4, n. 1, p. 64-69, 2015.

FELIPINI, R. B.; PIERO, R. M. D. Reducing the severity of apple bitter rot

in post-harvest by immersion of fruits in chitosan. **Pesquisa Agropecuâria Brasileira**, v. 44, n. 12, p. 1591-1597, 2009.

FERREIRA, D. F. **Sisvar: version 4.3**. Lavras: DEX/UFLA, 2003.

FILHO, A. B.; ALMEIDA, J. E. M.; LAMAS, C. Effect of Thiamethoxam on Entomopathogenic Microorganisms. **Neotropical Entomology**, v. 30, n. 3, p. 437447, 2001.

FILHO, I. A. P.; RODRIGUES, J. A. S. **Sorghum: the producer asks, Embrapa answers**. Brasilia, DF: Embrapa, 2015.

FILHO, L. C. P.; CIESLIK, L. F.; TALHEIMER, R.; LUCINI, M.; SILVEIRA, E. R. Mortality of anthills by different control methods in pastures.

Synergismus scyentifica **UTFPR**, v. 2, 2007.

FILHO, M. F. S.; COSTA, V. A.; PAZINI, W. C. **Manejo integrado de pragas na cultura da manga**, 2004. Available at: <https://www.researchgate.net/profile/Wilson_Pazini/publication/266493214_MANEJO_INTEGRADO_DE_PRAGAS_NA_CULTURA_DA_MANGA/links/58ac2865aca27206d9bf92e2/MANEJO-INTEGRADO-DE-PRAGAS-NA-CULTURA-DA MANGA.pdf> Accessed: 09-04-2018.

FILHO, W. R.; NICKELE, M. A.; STRAPASSON, P. **Combating leaf-cutting ants**. Curitiba: SENAR-PR, 2011a.

FILHO, W. R.; SANTOS, F.; STRAPASSON, P.; NICKELE, M. A. Damage caused by different levels of artificial defoliation to simulate the attack of leaf-cutting ants on Pinus taeda and Eucalyptus grandis. **Revista Pesquisa Florestal Brasileira**, v. 31, n. 65, p. 37-42, 2011b.

FILHO, W. R.; PORFiRIO-DA-SILVA, V.; NICKELE, M. A. MARTINS, M. F. O. **Leafcutter ants in integrated crop-livestock-forestry systems - iLPF: fundamentals for control**. 1. ed. Colombo, PR: Embrapa Florestas, 2013.

FILHO, W. R.; OLIVEIRA, E. **External activity, loading of granulated bait and control of** Acromyrmex crassispinus **in** Pinus taeda **forest**. 1. ed. Colombo, PR: Embrapa Florestas, 2002.

FINNEY, D. J. **Probit analysis**. 3. ed. Cambridge: Cambridge University Press, 1971.

FLORES, M. E. V. **Characterization of three strains of** Beauveria brongniartii **(Saccardo) Petch and their virulence on** Phthorimaea operculella **(Zeller) and** Symmetrischema tangolias **(Gyen)**. 2003. Tesis - Universidad Nacional Mayor de San Marcos, Lima, Peru.

FOLGARAIT, P.; GOROSITO, N.; POULSEN, M.; CURRIE, C. R. Preliminary In Vitro Insights into the Use of Natural Fungal Pathogens of Leaf-cutting Ants as Biocontrol Agents. **Current Microbiology**, v. 63, p. 250-258, 2011.

FORTI, L. C.; CAMARGO, R. S.; MATOS, C. A. O.; ANDRADE, A. P. P.; LOPES, J. F. Alloethism in Acromyrmex subterraneus brunneus Forel (Hymenoptera, Formicidae), during foraging, fungus garden cultivation and return of foraged materials. **Revista Brasileira de Entomologia**, v. 48, n. 1, p. 59-63, 2004.

FORTI, L. C.; NAGAMOTO, N. S.; RAMOS, V. M.; ANDRADE, A. P. P.; LOPES, J. F.

CAMARGO, R. S.; MOREIRA, A. A.; BOARETTO, M. A. C. Eficiencia de sulfluramida, fipronil y clorpirifos como sebos en el control de Atta capiguara Gonçalves (Hymenoptera: Formicidae). **Pasturas Tropicales**, v. 25, n. 3, p. 28-35, 2003.

FRANCE, A.; GERDING, M.; GERDING, M.; SANDOVAL, A. Pathogenicity of a collection of native strains of Metarhizium spp. and Beauveria spp. on Aegorhinus superciliosus, Asynonychus cervinus and Otiorhynchus sulcatus. **Agricultura Técnica**, v. 60, n. 3, 2000.

FRANCO, A. A.; PERES, A. R.; SOUZA, M. F. P.; QUEIROZ, M. S.; ASSIS, J. M. F. Action of plant extracts on the development of fungi symbionts of leaf-cutting ants. **Engenharia Ambiental**, v. 10, n. 1, p. 103-113, 2013.

FRATTI, I. G. B.; TORRES, M. A. P. 2014. **Efficiency in the transportation of toxic bark by leaf-cutting ants of the genus** Acromyrmex. Thesis - Faculty of Agronomy, University of the Republic, Montevideo, Uruguay.

GAJARDO, R.; PAVONE, D.; OSBORN, F.; DORTA, B. **Primer reporte de un hongo entomopatógeno sobre** Hylesia metabus **(Cramer, [1775]) (Lepidoptera: Saturniidae)**. In: HERNANDEZ, J. V.; OSBORN, F.; CONDE, J. E. Estudio multidisciplinario de la palometa peluda Hylesia metabus. Venezuela: Instituto Venezolano de Investigaciones Cientificas, 2012, p. 129-137.

GANDRA, L. C. **Mechanism of suppression of colonies of the leaf-cutting ant *Acromyrmex subterraneus subterraneus* by fipronil**. 2014. Dissertation (Postgraduate Diploma in Entomology) - Federal University of Viçosa, Viçosa, Minas Gerais.

GANDRA, L. C.; AMARAL, K. D.; COUCEIRO, J. C.; DELLA LUCIA, T. M. C.; GUEDES, R. N. C. Mechanism of leaf-cutting ant colony suppression by fipronil used in attractive toxic baits. **Pest Management Science**, v. 72, p. 1475-1481, 2016.

GANT, D. B.; CHALMERS, A. E.; WOLFF, M. A.; HOFFMAN, H. B.; BUSHEY, D. F. Fipronil: action at the GABA receptor. **Reviews in Toxicology**, v. 2, p. 147156, 1998.

GARCIA, M. A. G.; GARCIA, S. C.; GORDILLO, J. M. L.; MARTINEZ, R. F. M. Hongos entomopatógenos como uma alternativa en el Control Biológico. **Kuxulkab'**, v. 15, n. 27, p .25-28, 2008.

GASSEN, D. N.; TAMBASCO, F. J. Biological control of wheat aphids in Brazil. **Inf.**

Agropec., v. 9, n. 104, p. 49-51, 1983.

GAUGLER, R.; LEWIS, E.; STUART, R.J. Ecology in the service of biological control: the case of entomopathogenic nematodes. **Oecologia**, v. 109, p. 483-489, 1997.

GHIKAS, D. V.; KOUVELIS, V. N.; TYPAS, M. A. Phylogenetic and biogeographic implications inferred by mitochondrial intergenic region analyses and ITS1-5.8S- ITS2 of the entomopathogenic fungi Beauveria bassiana and B.brongniartii. **BMC Microbiology**, v. 10, 2010.

GHINI, R.; BETTIOL, W. Plant protection in sustainable agriculture. **Cadernos de Ciência & Tecnologia**, v. 17, n. 1, p. 61-70, 2000.

GIMENES, D. C.; ALEXANDRINO, T. D.; MACHADO, A. C.; VARÉA, G. S. Induction of Proteases from Beauveria bassiana by Aedes aegypti larvae and cuticle cicadas. **Biochemistry and Biotechnology Reports**, v. 3, n. 1, p. 41-47, 2014.

GOMES, M. D. F. **Insecticidal activity of entomopathogenic fungi and the alkaloid matrine on** Musca domestica **Linnaeus, 1758 (Diptera: Muscidae) larvae**. 2017. Dissertation (Post-Graduation in Tropical Medicine) - Federal University of Goiás, Goiânia.

GOMEZ, M. R.; CESAR, E. S.; RIVERA, J. S. M.; FLORES, J. L. R.; SALGADO, J. R. H.; MENDEZ, J. G. P. Evaluación de insecticidas alternativos para el control de paratrioza (Bactericera cockerelli B.y L.) (Homoptera: Triozidae) en el cultivo de chile jalapeno *(Capsicum annum* L.). **Revista Chapingo Serie Zonas Aridas**, v. 76, p. 47-56, 2008.

GÓMEZ, N. A. M. Hormigas cortadoras de hojas en el departamento del Vaupés, Colombia: Una propuesta de manejo integrado. **Revista Vaupés Innova**, p. 18-41, 2017.

GONZAGA, A. D.; SOUSA, S. G. A.; SILVA, N. M.; PEREIRA, J. O. Toxicity of Cow Urine and Cassava Manioc on Key Pests of Pineapple, **Rev. Bras. De Agroecologia**, v. 4, n. 2, 2009.

GONZALES, G. E. B. **Determination, biological cycle, biological parameters and behavior of** Telenomus **sp. (Hymenoptera: Scelionidae) parasitoid of eggs of** Liorhyssus hyalinus**. F (Hemiptera: Rhopalidae) "chinche de la quinua" under laboratory conditions, 2015-2016**. 2017. Tesis - Universidad Nacional de San Agustin de Arequipa, Arequipa, Peru.

GONZALEZ-CASTILLO, M.; AGUILAR, C. N.; RODRiGUEZ-HERRERA, R. Control of insect pests in agriculture using entomopathogenic weeds: challenges and perspectives. **Revista Cientifica de la Universidad Autònoma de Coahuila**, v. 4, n. 8, p. 42-55, 2012.

GROSMAN, D. M.; UPTON, W. W.; MCCOOK, F. A.; BILLINGS, R. F. Attractiveness and efficacy of fipronil and sulfluramid baits for control of the Texas leafcutting ant. **Southwestern Entomologist**, v. 27, n. 3/4, 2002.

GRUBER, A. K.; VALDIX, J. K. Control of *Atta* spp. with agricultural practices and botanical insecticides. **Manejo Integrado de Plagas y Agroecologia**, n. 67, p. 8790, 2003.

GUARANA, C. F. R. **Pathogenic action of strains of *Metarhizium anisopliae* on *Callosobruchus maculatus* (Coleoptera: Bruchidae), and chemical compatibility to insecticides**. 2007. Thesis (Post-Graduation in Biological Sciences) - Federal University of Pernambuco, Recife, Pernambuco.

GUÉDEZ, C.; CASTILLO, C.; CANIZALES, L.; OLIVAR, R. Biological control: a tool for sustainable and sustainable development. **ACADEMIA**, v. 7, n. 13, p. 50-74, 2008.

GUERRA, P. T.; WONG, L. J. G.; ROLDAN, H. M.; GUTIÉRREZ, C. G.;

PADILLA, C. R.; FLORES, R. A. G.; GUERRA, R. S. T. Bioinsecticidas: su empleo, producción

y comercialización en México. **Ciencia UANL**, v. 4, n. 2, p. 143-152, 2001.

GUERREIRO, J. C. The importance of ladybugs in the biological control of pests in Brazil and worldwide. **Revista Cientifica Eletrônica de Agronomia**, n. 5, 2004.

GUERRERO, J.; MERA, M.; SALVO, H.; CARRILLO, R. Discrimination of native Chilean strains of the long-lived entomopathogen Metarhizium anisopliae var. anisopliae with RAPD molecular markers. **Agro sur**, v. 28, n. 1, p. 81-91, 2000.

GUERRERO, J. R. **Potencial del botón de oro (**Tithonia diversifolia**) como controlador biológico de hormiga arriera,** Atta cephalotes **(Hymenoptera: Myrmicinae)**. 2013. Dissertation (Master's Degree in Biological Sciences) - Universidad Del Valle.

GUIMARÂES, A. G. L. P. **Production of** conidia **and hydrolytic enzymes by** Beauveria bassiana **(Bals) Vuillemin (Deuteromycotina: Hyphomycetes) in different substrates**. 2016. Dissertation (Post-Graduation in Biotechnology) - Federal University of Paraiba, Joao

Pessoa, Paraiba.

GUIMARÂES, M. R. F. **Polyethism and life expectancy in *Atta laevigata* workers**. 2010. Dissertation (Post-Graduation in Entomology) - Federal University of Viçosa, Viçosa, Minas Gerais.

GUSMÂO, L. F. P.; GRANDI, R. A. P. Hyphomycetes with conidiomata of the sporodochium and sinema types associated with leaves of Cedrela fissilis (Meliaceae), in Maringà, PR, Brazil. **Acta Botanica Brasilica**, v. 11, n. 2, p. 123-134, 1997.

GUTIÉRREZ, A. I.; SALDARRIAGA, Y. Observation of the pathogenicity of Metarhizium anisopliae in Nasutitermes sp. soldiers (Isoptera: Termitidae). **Revista Colombiana de Entomologia**, v. 30, n. 2, p. 151-156, 2004.

GUTIÉRREZ, M. E. M.; CARDENAS, Y. G.; LEQUERQUE, A. El género Metarhizium Sorokin y su aplicación en el control de insectos plagas. **Revista Cubana de Ciencias Biológicas**, v. 5, n. 2, p. 1-13, 2016-2017.

HAEDER, S.; WIRTH, R.; HERZ, H.; SPITELLER, D. Candicidin-producing Streptomyces support leaf-cutting ants to protect their fungus garden against the pathogenic fungus Escovopsis. **Proceedings of the National Academy of Sciences**, v. 106, n. 12, p. 4742-4746, 2009.

HAINES, B. L. Element and energy flows through colonies of the leaf cutting ant Atta colombica, in Panama. **Biotropica**, v. 10, n. 4, p. 270-277, 1978.

HANSON, P.; HILJE, L. **Control Biologico de Insectos**. Turrialba: Centro Agronomico Tropical de Investigacion y Ensenanza, 1993.

HIDALGO, E. **Control de plagas agricolas y Aorestales con agentes microbiológicos**. Turrialba, Costa Rica: CATIE, 1999.

HOFFMANN-CAMPO, C. B.; MOSCARDI, F.; CORRÊA- FERREIRA, B. S.;

OLIVEIRA, L. J.; SOSA-GÓMEZ, D. R.; PANIZZI, A. R.; CORSO, I. C.;

GAZZONI, D. L.; OLIVEIRA, E. B. **Soybean pests in Brazil and their integrated management**. Londrina: Embrapa Soja, 2000.

HOLLDOBLER, B.; WILSON, E. O. **The Ants**. Cambridge: Harvard University Press, 1990.

HUANES-CARRANZA, J.; WILSON-KRUGG, J.; Efecto de Beauveria bassiana y

Metarhizium anisopliae sobre adultos y ninfas de Oligonychus sp. en condiciones de laboratório. **REBIOL**, v. 36, n. 1, p. 51-58, 2016.

HUFFAKER, C. B. **Biological control**. Boston, 1969.

HUMBER, R. A. **Identification of entomopathogenic fungi**. In: LACEY, L. A.

Manual of techniques in invertebrate pathology. 2. ed. Academic Press / Elsevier, 2012, p. 151-187.

IKEDA, T.; NAGATA, K.; KONO, Y.; YEH, J. Z.; NARAHASHI, T. Fipronil modulation of GABAA receptor single-channel currents. **Pest Management Science**, v. 60, p. 487-492, 2004.

ITO, E. T.; VARÉA-PEREIRA, G.; MIYAGUI, D. T.; PINOTTI, M. H. P.; NEVES, P. M. O. J. Production of Extracellular Protease by a Brazilian Strain of Beauveria bassiana Reactivated on Coffee Berry Borer, Hypothenemus hampei. **Brazilian archives of**

Biology and Technology, v. 50, n. 2, p. 217-223, 2007.

JACKSON, M. A.; PAYNE, A. R.; ODELSON, D. A. Liquid-culture production of blastospores of the bioinsecticidal fungus Paecilomyces fumosoroseus using portable fermentation equipment. **Journal of Industrial Microbiology and Biotechnology**, v. 31, p. 149-154, 2004.

JONKMAN, J. C. M. **Biology and ecology of the leaf-cutting ant** Atta vollenweideri

Forel, 1898 (Hymenoptera: Formicidae) and its impact in Paraguayan pastures. Lien, 1977.

JORGE, M. L. A. **Characterization of Portuguese isolates of *Albugo candida* (Pers.) Kuntze**. 1998. Dissertation (Master's Degree in Horticulture, Fruticulture, Olericulture and Ornamental Plants) - Technical University of Lisbon, Lisbon, Portugal.

JUNIOR, A. G.; BIZI, R. M.; AUER, C. G. ***Oidium* sp. in Eucalyptus Seedlings**. 1. ed. Colombo, PR: Embrapa Florestas, 2005.

JÛNIOR, A. G.; SANTOS, A. F.; AUER, C. G. Perspectives on the use of biological control against forest diseases. **Floresta**, v. 30, n. 1/2, p. 155-165, 2000.

JÛNIOR, A. L. M.; DELLA LUCIA, T. M. C.; BARBOSA. L. C. A.; MAFFIA, L. A.; MORANDI, M. A. B. Effect of Secretions from the Mandibular Gland of *Atta sexdens rubropilosa* Forel (Hymenoptera: Formicidae) on the Germination of Conidia of *Botrytis cinerea* Pers. **Neotropical Entomology**, v. 30, n. 3, p. 403406, 2001.

JÛNIOR, G. M. C. **Integrated analysis of the virulence and conidia production variables in the selection of entomopathogenic fungi for the development of biopesticides**. 2017. Dissertation (Master of Science) - University of São Paulo, Piracicaba.

JÛNIOR, F. B. P. **Pathogenic action of *Beauveria bassiana* strains on *Callosobruchus maculatus* (COLEOPTERA: BRUCHIDAE), genetic analysis (PCR) and compatibility with chemical insecticides**. 2006. Thesis (Post-Graduation in Fungal Biology) - Federal University of Pernambuco, Recife, Pernambuco.

KAGIMURA, F. Y.; KASBURG, C. R.; FREITAS, D.; VILANI, A.; QUEIRÓZ, J. A.; DAMIN, S.; ONOFRE, S. B. Action of fungicides on the growth of the entomopathogenic fungus *Beauveria bassiana* (Bals.) Vuill. **Revista em Agronegócios e Meio Ambiente**, v. 4, n. 2, p. 277-290, 2011.

KESSLER. P.; ENKERLI, J.; SCHWEIZER, C.; KELLER, S. Survival of *Beauveria brongniartii* in the soil after application as a biocontrol agent against the European cockchafer *Melolontha melolontha*. **Biocontrol**, v. 49, p. 563-581, 2004.

KLEESPIES, R. G.; ZIMMERMANN, G. Effect of additives on the production, viability and virulence of blastospores of Metarhizium anisopliae. **Bioc. Sci. Tech.**, v. 8, p. 207-214, 1998.

KLEINEIDAM, C.; ERNST, R.; ROCES, F. Wind-induced ventilation of the giant nests of the leaf-cutting ant Atta vollenweideri. **Naturwissenschaften**, v. 88, p. 301-305, 2001.

KLICH, M. A.; SAMSON, R. A. Aspergillus **reference cultures**. International Union of Microbiological Societies, 1996.

KOST, C.; LAKATOS, T.; BOTTCHER, I.; ARENDHOLZ, W.; REDENBACH, M.; WIRTH, R. Non-specific association between filamentous bacteria and fungusgrowing ants. **Naturwissenschaften**, v. 94, p. 821-828, 2007.

LA, M. T. D.; CORTEZ, H.; ORTIZ, C. F.; CAPPELLO, S.; CRUZ, A. D. L. Caracterización de aislamientos nativos de Metarhizium anisopliae y su patogenicidad hacia Aeneolamia postica, en Tabasco, México. **Revista Colombiana de Entomologia**, v. 39, n. 1, p. 40-46, 2013.

LAURENTIS, V. L. Helicoverpa armigera **(HÜBNER) (Lepidoptera: Noctuidae): tactics for integrated management**. 2017. Thesis (Doctorate in Agronomy) - Universidade Estadual Paulista, Jaboticabal.

LAZO, M. L. S. R. **Characterization and pathogenicity of entomopathogenic fungi isolated from the eucalyptus tan bug, *Thaumastocorisperegrinus* (Hemiptera: Thaumastocoridae)**. 2012. Dissertation (Master's Degree in Agronomy) - Universidade Estadual Paulista "Jûlio de Mesquita Filho", Botucatu, Sao Paulo.

LEAL, I. R.; WIRTH, R.; TABARELLI, M. The Multiple Impacts of Leaf-Cutting Ants and Their Novel Ecological Role in Human-Modified Neotropical Forests. **Biotropica**, v. 46, n. 5, p. 516-528, 2014.

LECUONA, R. E.; RODRIGUEZ, J.; ROSSA, F. R. L. Effect of Constant and Cyclical Temperatures on the Mortality of Triatoma infestans (Klug) (Hemiptera: Reduviidae) Treated with Beauvaria bassiana (Bals.) Vuill. (Hyphomycetes). **Neotropical Entomology**, v. 34, n. 4, p. 675-679, 2005.

LEITE, L. G.; BATISTA FILHO, A.; ALMEIDA, J. E. M.; ALVES, S. B. **Produçâo de fungos entomopatogênicos**. Ribeirao Preto: Livroceres, 2003.

LEITE, M. S. P.; IEDE, E. T.; PENTEADO, S. R. C.; ZALESKI, S. R. M.;

CAMARGO, J. M. M.; RIBEIRO, R. D. Selection of isolates of entomopathogenic fungi for the control of *Hedypathes betulinus* and evaluation of persistence. **Floresta**, v. 41, n. 3, p. 619-628, 2011.

LEMUS, Y. A.; RODRiGUEZ, G. M.; CUERVO, R. A.; VANEGAS, J. A. D.; ZULUAGA, C. L.; RODRiGUEZ, G. Determination of the feasibility of *Metarhizium anisopliae* to be used as a biological control of the fire ant *(Atta cephalotes)*. **Revista Cientifica Guillermo de Ockham**, v. 6, n. 1, p. 91-98, 2008.

LENTEREN, J. C.; WOETS, J. Biological and integrated pest control in greenhouses. **Annual Review of Entomology**, v. 33, p. 239-269, 1988.

LEÓN, G. A. Insect diversity in citrus and its importance in integrated pest management programs. **Manejo Integrado de Plagas y Agroecologia**, n. 74, p. 85-93, 2005.

LERMA, J. M.; ULLOA, P. C.; MANZANO, M. R. Caracterización de nidos de la hormiga arriera *Atta cephalotes* (Hymenoptera: Myrmicinae) en Cali (Colombia).

Revista Colombiana de Entomologia, v. 32, n. 2, p. 151-158, 2006.

LIMA, D. B. C.; SILVA, A. G.; PROCÓPIO, S. O.; BARROSO, A. L. L.; DAN, H. A. Chemical control of volunteer Roundup Ready® soybean plants at different stages of development. **Revista Caatinga**, v. 24, n. 3, p. 64-70, 2011.

LINK, H. M.; LINK, F. M.; LINK, D. Control of the black baker ant, *Acromyrmex crassispinus*, with powdered formicides. **Ciência Floretal**, v. 10, n. 1, p. 45-56, 2000.

LIRA, D. D. **Characterization of *Aspergillus* sp. in terms of their ability to degrade diesel oil**. 2014. Dissertation (Postgraduate Diploma in Fungal Biology) - Federal University of Pernambuco, Recife, Pernambuco.

LOECK, A. E.; GRUTZMACHER, D. D.; COIMBRA, S. M. Occurrence of leaf-cutting ants of the genus *Acromyrmex* in the main agricultural regions of Rio Grande do Sul. **Revista Brasileira de Agrociência**, v. 9, n. 2, p. 129-133, 2003.

LOECK, A. E.; GRUTZMACHER, D. D.; STORCH, G. Geographical distribution of *Atta sexdenspiriventris* Santschi, 1919, in the main agricultural regions of the state of Rio Grande do Sul. **Revista Brasileira de Agrociência**, v. 7 n. 1, p. 54-57, 2001.

LOPES, A. J.; JANSEN, U.; CAPONE, D.; JANSEN, J. M. Aspergiloses pulmonares. **Pulmao**, v. 13, n. 1, p. 34-44, 2004.

LOPES, R. S.; PORTELA, A. P. A. S.; SVEDESE, V. M.; ALBUQUERQUE, A. C.; LIMA, E. A. L. A. Morphological aspects of Paecilomyces farinosus (HOLM EX S. F. GRAY)

BROWN & SMITH on infection in *Coptotermes gestroi* (WASMANN) (ISOPTERA: RHINOTERMITIDAE). **Biological**, v. 70, n. 1, p. 29-33, 2008.

LOPES, R. S. **Pathogenicity of** Paecilomyces farinosus **on** Coptotermes ***gestroi and* biological parameters**. 2007. Dissertation (Post-graduation in Fungal Biology) - Federal University of Pernambuco, Recife, Pernambuco.

LOPES, R. S.; LIMA, G.; OLIVEIRA, L. G.; COSTA, A. F.; CORREIA, M. T. S.; LIMA, E. A. L. A.; LIMA, V. L. M. Biological aspects of *Isaria* species (Persoon) after infection on *Sitophilus zeamais* Motschulsky (Coleoptera: Curculionidae). **Pesq. agropec.**, v. 21, n. 1, p. 32-38, 2016.

LOUREIRO, E. S.; FILHO, A. B.; ALMEIDA, J. E. M.; PESSOA, L. G. A. Selection of

Isolates of *Metarhizium anisopliae* (Metsch.) Sorok. Against the Sugarcane Root Cicada *Mahanarva fimbriolata* (Stâl) (Hemiptera: Cercopidae) in the Laboratory. **Neotropical Entomology**, v. 34, n. 5, p. 791-798, 2005a.

LOUREIRO, E. S.; FILHO, A. B.; ALMEIDA, J. E. M.; PESSOA, L. G. A.

Production of isolates of *Metarhizium anisopliae*, selected for the control of *Mahanarva fimbriolata* (Stal, 1854). **Arquivos do Instituto Biológico**, v. 72, n. 4, p. 469-472, 2005b.

LOUREIRO, E. S.; MOINO JR., A.; ARNOSTI, A.; SOUZA, G. C. Effect of chemical phytosanitary products used on lettuce and chrysanthemum on entomopathogenic fungi. **Neotropical Entomology**, v. 31, n. 2, p. 263-269, 2002.

LOUREIRO, E. S.; MONTEIRO, A. C. Pathogenicity of isolates of three entomopathogenic fungi to soldiers of *Atta sexdens sexdens* (Linnaeus, 1758) (Hymenoptera: Formicidae). **Revista Arvore**, Viçosa, Minas Gerais, v. 29, n. 4, p. 553561, 2005.

LOURENÇO, C. T.; CARVALHO, S. M.; MALASPINA, O.; NOCELLI, R. C. F. Oral toxicity of Fipronil insecticide against the stingless bee *Melipona scutellaris* (Latreille, 1811). **Bulletin of Environmental Contamination and Toxicology**, v. 89, p. 921-924, 2012.

LUNA ALVES-LIMA, E. A.; TIGANO, M. S. Cytology of yeast-like structures of *Beauveria bassiana* in Iiquid culture media and in the hemolymph of *Spodoptera furngiperda*. **Revista de Microbiologia**, v. 20, p. 85-94, 1989.

Las hormigas "arrieras", *Atta* spp. (Hymenoptera: Formicidae) de México. **Dugesiana**, v. 3, n. 1, p. 33-35, 1996.

MADELIN, M. F. **Diseases caused by Hyphomycetous fungi**. In: STEINHAUS, E. A. Insect Pathology: An Advanced Treatise. New York: Academic Press, 1963, p. 233-271.

MARANHÂO, E. A. A.; MARANHÂO, E. H. A. Hongos entomopatógenos: importante herramienta para el control de "moscas blancas" (Homoptera: Aleyrodidae). **Anais da Academia Pernambucana de Ciência Agronômica**, v. 5/6, p. 209-242, 2008-2009.

MARiN, P. M.; PARDEY, A. E. B. Selección de aislamientos de *Beauveria bassiana* virulentos a *Compsus* n. sp. **Cenicafé**, v. 59, n. 2, p. 165-173, 2008.

MARQUES, A.; DIONÎSIO, L.; PALLERO-BUENO, F.; NETO, L. Study of the entomopathogenic capacity of phytopathogenic fungi. **Actas Portuguesas de Horticultura**,

n. 25, p. 114-119, 2016.

MARTINEZ, M. J. G. **Evaluación de la patogenicidad y esporulación del hongo** Metarhizium anisopliae **(Metsch.) Sorokin cepa Metagreen en concentración de 10^{10} c/ml sobre adulto de Mosca Blanca** (Bemisia tabaci) **en el Laboratorio de Hongos Entomopatógenos del Campus Agropecuario de UNAN-León 2011-2012**. 2014. Thesis - Universidad Nacional Autónoma de Nicaragua, León, Nicaragua.

MASCARIN, G. M.; JARONSKI, S. T. The production and uses of *Beauveria bassiana* as a microbial insecticide. **World Journal of Microbiology and Biotechnology**, v. 32, 2016.

MASON, S. C.; SGARBI, C.; COVACHINA, J. C.; PENA, J. M.; BERENSZTEIN, N. D.; MARGARÎA, C.; RICCI, M. *Acromyrmex* Mayr (Hymenoptera: Formicidae: Myrmicinae): patterns of species distribution in the province of Buenos Aires, Argentina. **Revista del Museo Argentino de Ciencias Naturales**, v. 19, n. 2, p. 185-199, 2017.

MASUTI, C.S.M; MERMUT, A.R. 2007. Sorption of fipronil and its sulfide derivative by soils and goethite. Geoderma 140 (1-2) 01-07.

MATTE, W. D.; QUEIROZ, L. F.; CORASSA, J. N. First record of *Atta sexdens rubropilosa* Forel, 1908 (Hymenoptera: Formicidae) attacking noni (*Morinda* citrifolia L.) in Colorado do Oeste (RO), Brazil. **Arquivos do Instituto Biológico**, v. 83, p. 1-4, 2016.

MATTOSO, T. C. **Role of ectosymbiont bacteria in** Acromyrmex ***subterraneus subterraneus*** **(Forel, 1901) in protection against entomopathogenic fungi**. 2012. Dissertation (Master's Degree in Plant Production) - Universidade Estadual do Norte Fluminense Darcy Ribeiro, Campos dos Goytacazes, Rio de Janeiro.

MASIULIONIS, V. E. **Fungi associated with** Acromyrmex **and basal** Attini **ants from Argentina and Brazil**. 2013. Thesis (Doctorate in Biological Sciences) - Universidade Estadual Paulista "Julio de Mesquita Filho", Rio Claro.

MAYERHOFER, J.; ENKERLI, J.; ZELGER, R.; STRASSER, H. Biological control of the European cockchafer: persistence of Beauveria brongniartii after long-term applications in the Euroregion Tyrol. **Biocontrol**, v. 60, p. 617-629, 2015.

MELO, D. R.; CRUZ, G. B.; REIS, R. C. S.; BITTENCOURT, V. R. E. P. Development of the fungi Metarhizium anisopliae (Metschnikoff, 1879) Sorokin, 1883 and Beauveria bassiana (Balsamo) Vuillemin, 1912 on Ctenocephalides felis *felis* (Bouché, 1835). **Revista Brasileira**

de Parasitologia Veterinària, v. 16, n. 3, p. 166-170, 2007.

MERTZ, N. R.; ALVES, L. F. A.; MARCOMINI, A. M.; OLIVEIRA, D. G. P.;

SANTOS, J. C. Effect of natural phytosanitary products on *Beauveria bassiana* (Bals.) Vuill. *in vitro*. **BioAssay**, v. 5, n. 3, p. 1-10, 2010.

MEYER, S. T.; NEUBAUER, M.; SAYER, E. J.; LEAL, I. R.; TABARELLI, M.; WIRTH, R. Leaf-cutting ants as ecosystem engineers: topsoil and litter disturbances around *Atta cephalotes* nests reduce nutrient availability. **Ecological Entomology**, v. 38, p. 497-504, 2013.

MINTER, D. W.; HAWKSWORTH, D. L.; ONIONS, A. H. S.; KOZAKIEWICZ, Z. **Descriptive terminology of the conidiogenous structures in** Aspergillus **and** Penicillium. In: SAMSON, S. A.; PITT, J. I. Advances in *Penicillium* and *Aspergillus* systematics. New York: Plenum Press, 1985.

MIRANDA, J. E.; ARAUJO, L. H. A. **Pests of the sesame crop: biology, damage and control methods**. 1. ed. Campina Grande, Paraiba: Embrapa Algodao, 2003.

MIYASHIRA, C. H. **Influence of caffeine on the survival of sauviae *Atta sexdens rubropilosa* (Hymenoptera: Fomicidae) and on the *in vitro* growth of its**

mutualistic fungus. 2007. Dissertation (Master of Science) - University of Sao Paulo, Sao Paulo.

MOINO JR., A.; ALVES, S. B. Determination of Concentrations of *Beauveria bassiana* (Bals.) Vuill. for the Control of Insect Pests of Stored Grain. **Anais da Sociedade Entomológica do Brasil**, v. 26, n. l, p. 15-20, 1997.

MOINO JR., A.; ALVES, S. B. Efeito de imidacloprid e fipronil sobre *Beauveria bassiana* (Bals.) Vuill. e *Metarhizium anisopliae* (Metsch.) Sorok. e no comportamento de limpeza de *Heterotermes tenuis* (Hagen), **Anais da Sociedade Entomológica do Brasil**, v. 27, n. 4, p. 611-619, 1998.

MOINO JR., A.; ALVES, S. B.; LOPES, R. B.; NEVES, P. M. O. J.; PEREIRA, R. M.; VIEIRA, S. A. External development of the entomopathogenic fungi *Beauveria bassiana* and *Metarhizium anisopliae* in the subterranean termite *Heterotermes tenuis*. **Scientia Agricola**, v. 59, n. 2, p. 267-273, 2002.

MONZÓN, A. Producción, uso y control de calidad de hongos entomopatógenos en Nicaragua. **Manejo Integrado de Plagas**, n. 63, p. 95-103, 2001.

MONTOYA-CORREA, M.; MONTOYA-LERMA, J.; ARMBRECHT, I. ROPERO, M. C. G. Cómo responde la hormiga cortadora de hojas *Atta cephalotes* (Hymenoptera: Myrmicinae) a la remoción mecânica de sus nidos? **Boletin del Museo de Entomologia de la Universidad del Valle**, v. 8, n. 2, p. 1-8, 2007.

MORA, M. A. E.; CASTILHO, A. M. C.; FRAGA, M. E. Entomopathogenic fungi: enzymes, toxins and factors affecting diversity. **Revista Brasileira de Produtos Agroindustriais**, v. 18, n. 3, p. 335-349, 2016.

MORAES, A. M. L.; COSTA, G. L.; BARCELLOS, M. Z. C.; OLIVEIRA, R. L.; OLIVEIRA, P. C. The entomopathogenic potential of *Aspergillus* spp. in mosquitoes vectors of tropical diseases. **Journal of Basic Microbiology**, v. 41, n. 1, p. 45-49, 2001.

MOREIRA, A. A.; FORTI, L. C.; CASTELLANI, M. A.; ANDRADE, A. P. P. Nest architecture of grass-cutting ants. **Biológico**, v. 69, Supl. 2, p. 83-85, 2007.

MOREIRA, R. A. SANTOS, R. A.; ROCHA, O. Preliminary assessments of the acute toxicity of the pesticide fipronil to the rotifers *Asplanchna sieboldi, Brachionus caudatus* and *Lepadella patella*. **Fórum Ambiental**, v. 8, n. 12, p. 63-75, 2012.

MORESSI, M.; NETO, A. M.; CREPALDI, R. A.; CARBONARI, V.; DEMÉTRIO, M. F.; SILVESTRE, R. Efficiency of mechanical control of leaf-cutting ants

(Atta laevigata) in reforestation with native species. **Biological**, v. 69, Suppl. 2, p. 471-473, 2007.

MOUTINHO, P.; NEPSTAD, D. C.; DAVIDSON, E. A. Influence of leaf-cutting ant nests on secondary forest growth and soil properties in Amazonia. **Ecology**, v. 84, n. 5, p. 1265-1276, 2003.

NAVA, D. E.; ZANARDI, O. Z.; MELO, M.; SILVA, S. D. A. **Pest and beneficial insects in tung cultivation**. Pelotas: Embrapa Clima Temperado, 2009.

NEDER, R. N. **Microbiologia: Manual de laboratòrio**. Sao Paulo: Nobel, 1992.

NETO, J. A. D. **Association and compatibility of chemical products and the fungi** Trichoderma harzianum **and** Paecilomyces lilacinus **in the management of**

phytonematodes in soybean crops. 2014. Dissertation (Master's Degree in Agronomy) - Federal University of Mato Grosso do Sul, Chapadao do Sul, Mato Grosso do Sul.

NEVES, P. M. O. J.; ALVES, S. B. External Events Related to the Infection Process of Cornitermes cumulans (Kollar) (Isoptera: Termitidae) by the Entomopathogenic Fungi Beauveria bassiana and Metarhizium anisopliae. **Neotropical Entomology**, v. 33, n. 1, p. 51-56, 2004.

NEVES, P. M. O. J.; HIROSE, E.; TCHUJO, P. T.; MOINO JR, A. Compatibility of Entomopathogenic Fungi with Neonicotinoid Insecticides. **Neotropical Entomology**, v. 30, n. 2, p. 263-268, 2001.

OKUMURA, F. **Voltammetric study and development of electroanalytical methodology for the determination of the insecticide fipronil using the graphite-polyurethane composite electrode**. 2009. Thesis (Post-graduation in Chemistry), University of Sao Paulo, Sao Carlos.

OLIVEIRA, A. M.; MARACAJÂ, P. B.; FILHO, E. T. D.; LINHARES, P. C. F. Biological control of pests in commercial crops as an alternative to the use of pesticides. **Revista Verde**, v. 1, n. 2, p. 1-9, 2006.

OLIVEIRA, B. M. S.; MELO, C. R.; ALVES, P. B.; SANTOS, A. A.; SANTOS, A. C. C.; SANTANA, A. S.; ARAÙJO, A. P. A.; NASCIMENTO, P. E. S.; BLANK, A. F.; BACCI, L. Essential Oil of Aristolochia trilobata: Synthesis, Routes of Exposure, Acute Toxicity, Binary Mixtures and Behavioral Effects on Leaf-Cutting Ants. **Molecules**, v. 22, 2017.

OLIVEIRA, C. N.; NEVES, P. M. O. L.; KAWAZOE, L. S. Compatibility between the entomopathogenic fungus Beauveria bassiana and insecticides used in coffee plantations. **Scientia Agricola**, v. 60, n. 4, p. 663-667, 2003.

OLIVEIRA, F. R. **Prospecting fungi for the control of *Anticarsia gemmatalis* HÜBNER, 1818 (Lepidoptera: Noctuidae)**. 2013. Dissertation (Post-graduation in Agrobiology) - Federal University of Santa Maria, Santa Maria, Rio Grande do Sul.

OLIVEIRA, R. C.; ALVES, L. F. A.; NEVES, P. M. O. J. Susceptibility of Oligonychus yothersi (Acari: Tetranychidae) to the fungus Beauveria bassiana. **Scientia Agricola**, v. 59, n. 1, p. 187-189, 2002a.

OLIVEIRA, R. C.; NEVES, P. M. O. J.; GUZZO, E. C.; ALVES, V. S. Compatibility of entomopathogenic fungi with agrochemicals. **Semina: Ciências Agrârias**, v. 23, n. 2, p. 211-216, 2002b.

OLIVEIRA, R. C; NEVES, P. M. O. J.; ALVES, L. F. A. Seleçao de Fungos Entomopatogênicos Para o Controle de Oligonychus yothersi (McGregor) (Acari: Tetranychidae), na Cultura da Erva-Mate (Ilex paraguariensis St. Hill.). **Neotropical Entomology**, v. 33, n. 3, p. 347-351, 2004.

ORTIZ, A. G.; FILHO, O. P.; SANTOS, A.; SOUZA, M. D.; FAVARES, L. G.; NASCIMENTO, D. A. Foraging response of Acromyrmex rugosus (Smith, 1858) and Acromyrmex balzani (Emery, 1890) (Hymenoptera: Formicidae) to *Eucalyptus Camaldulensis* Dehnh. seedlings with different nutritional restrictions. **Revista Espacios**, v. 38, n. 44, 2017.

OTTATI-DE-LIMA, E. L. **Production of *Metarhizium anisopliae* (METSCH.) Sorok. and *Beauveria bassiana* (BALS.) Vuill. in different substrates and the effect of ultraviolet radiation and temperature on the infective structures of these entomopathogens**. 2007. Thesis (Doctorate in Agronomy) - Universidade Estadual Paulista "Jûlio Mesquita Filho", Botucatu, Sao Paulo.

OTTATI-DE-LIMA, E. L.; FILHO, A. B.; ALMEIDA, J. E. M.; GASSEN, M. H.; WENZEL, I. M.; ALMEIDA, A. M. B.; ZAPELLINI, L. O. Semisolid production of *Metarhizium anisopliae* and *Beauveria bassiana* in different substrates and the effect of ultraviolet radiation and temperature on propagules of these entomopathogens.

Arquivos do Instituto Biológico, v. 77, n. 4, p. 651-659, 2010.

PACOLA-MEIRELLES, L. D.; AZEVEDO, J. L.Natural variability of the entomopathogenic fungus *Beauveria bassiana.* **Arquivos de Biologia e Tecnologia**, v. 33, n. 3, p. 657-672, 1990.

PADULLA, L. F. L.; ALVES, S. B. Susceptibility of Diaphorina citri nymphs to entomopathogenic fungi. **Arquivos do Instituto Biológico**, v. 76, n. 2, p. 297-302, 2009.

PALMA, A. M. C. **Biological baits for the control of leaf-cutting ants (Hymenoptera: Formicidae)**. 2016. Monograph (Forest Engineering Course) - University of Recôncavo da Bahia, Cruz das Almas, 2016.

PARI, J. G. Z.; ROJAS, P. V.; LANDAETA, J. C. Z.; MAMANI, F. E. Evaluación de Substratos y Producción de Beauveria brongniartii (Sacc.) Petch para Control de Gorgojo de los Andes (premnotrypes spp) en Cultivo de Papa. **Revista de Investigaciones Altoandina**, v. 17, n. 3, p. 347-354, 2015.

PASTOR, A. R. **Control of chestnut carpophagous insects *(Castanea sativa)* in Andalusia through mass capture with sex pheromone and evaluation of the insecticidal activity of entomopathogenic weeds**. 2013. Thesis (Doctorate) - Universidad de Córdoba, Córdoba.

PAULI, G.; LOPES, R. B.; ALVES, S. B.; JUNIOR, E. R. D.; MASCARIN, G. M. False borer increases dissemination of *Beauveria bassiana* in field populations of

banana rhizome borer. **Ciência Rural**, v. 41, n. 11, p. 1867-1870, 2011.

PAVONE, D.; DORTA, B. Efecto de agroquimicos sobre el desarrollo del hongo entomopatógeno *Nomuraea rileyi* y su virulencia sobre *Spodoptera frugiperda*.

Bioagro, v. 22, n. 2, 2010.

PEDRINI, N.; ORTIZ-URQUIZA, A.; HUARTE-BONNET, C.; ZHANG, S.;

KEYHANI, N. O. Targeting of insect epicuticular lipids by the entomopathogenic fungus *Beauveriabassiana*: hydrocarbon oxidation within the context of ahost- pathogen interaction. **Frontiers in Microbiology**, v. 4, n. 24, p. 1-18, 2013.

PÉREZ, A. C.; ELÓSEGUI, O.; PADRÓN, N. B. Aislamiento, caracterización morfológica y fisiológica del hongo entomopatógeno *Paecilomyces fumosoroseus* (WIZE) Broum & Smith. **Fitosanidad**, v. 7, n. 3, p. 27-32, 2003.

PERRI, D.; GOROSITO, N.; FERNANDEZ, P.; BUTELER, M. Plant-based compounds with potential as push-pull stimuli to manage behavior of leaf-cutting ants. **Entomologia Experimentalis et Applicata**, v. 163, p. 150-159, 2017.

PIATTI, P.; CRAVANZOLA, F.; BRIDGE, P. D.; OZINO, O. I. Molecular characterization of *Beauveria brongniartii* isolates obtained from *Melolontha melolontha* in Valle d'Aosta (Italy) by RAPD-PCR. **Letters in Applied Microbiology**, v. 26, p. 317-324, 1998.

PIMENTA, L. B.; ARAÙJO, M. S.; LIMA, R.; SILVA, J. M. S.; NAVES, V. G. O. Foraging dynamics and colony characterization of *Acromyrmex balzani* (Emery, 1890) (Hymenoptera: Formicidae) in a cerrado environment in Goiás. **Revista Cientifica Eletrônica de**

Engenharia Florestal, n. 9, 2007.

PINTO, A. P. F. **Pathogenicity of entomopathogenic fungi and compatibility with insecticides for the management of** Diaphorina citri **Kuwayama, (Hemiptera: Liviidae)**. 2016. Thesis (Doctorate in Agronomy). Universidade Estadual Paulista "Júlio de Mesquita Filho", Botucatu, Sao Paulo.

POTRICH, M.; ALVES, L. F. A.; HAAS, J.; SILVA, E. R. L.; DAROS, A.; PIETROWSKI, V.; NEVES, P. M. O. J. Selectivity of *Beauveria bassiana* and *Metarhizium anisopliae* to *Trichogramma pretiosum* Riley (Hymenoptera: Trichogrammatidae). **Neotropical Entomology**, v. 38, n. 6, p. 822-826, 2009.

POULSEN, M.; BOT, A. N. M.; CURRIE, C. R.; BOOMSMA, J. J. Mutualistic bacteria and a possible trade-off between alternative defense mechanisms in *Acromyrmex* leaf-cutting ants. **Insectes Sociaux**, v. 49, p. 15-19, 2002.

POULSEN, M.; HUGHES, W. O. H.; BOOMSMA, J. J. Differential resistance and the importance of antibiotic production in *Acromyrmex echinatior* leaf-cutting ant castes towards the entomopathogenic fungus *Aspergillus nomius*. **Insectes Sociaux**, v. 53, p. 349-355, 2006.

QUESADA-MORAGA, E.; RUIZ-GARCIA, A.; SANTIAGO-ALVAREZ, C. Laboratory Evaluation of Entomopathogenic Fungi *Beauveria bassiana* and *Metarhizium anisopliae*

Against Puparia and Adults of *Ceratitis capitata* (Diptera: Tephritidae). **Journal of Economic Entomology**, v. 99, n. 6, p. 1955-1966, 2006.

RAMiREZ, H. G.; GRANJA, A. Z.; AGUILA, E. T. D.; CANTORAL, M. T.

Manual

production and use of entomopathogenic weeds. Peru: Servicio Nacional de Sanidad Agraria, 2014.

RAMOS, E. Q.; ALVES, S. B.; DEMÉTRIO, C. G. B.; COSTA, S. C. Selection of entomopathogenic fungi for the control of *Bemisia tabaci* biotype B. **Manejo Integrado de Plagas y Agroecologia**, n. 73, p. 21-28, 2004.

RAMOS, L. S.; MARINHO, C. G. S.; ZANETTI, R.; DELABIE, J. H. C.; SCHLINDWEIN, M. N. Impact of Granulated Formicide Baits on Non-Target Myrmecofauna in Eucalyptus Trees According to Two Application Methods. **Neotropical Entomology**, v. 32, n. 2, p. 231-

237, 2003.

RAMOS, V. M.; FORTI, L. C.; BOARETTO, M. A. C.; MOREIRA, A. A.; LOPES, J. F. S. Attractiveness of citrus pulp baits sprayed with jaraguà grass extract *(Hyparrhenia rufa* Nees) to the grass-cutting ant *Atta capiguara*. **Pasturas Tropicales**, v. 28, n. 3, 2006.

RANDO, J. S. S.; FORTI, L. C. Occurrence of *Acromyrmex* ants Mayr, 1865, in some municipalities in Brazil. **Acta Scientiarum. Biological Sciences**, v. 27, n. 2, p. 129-133, 2005.

RASHID, M.; GARJAN, A. S.; NASERI, B.; GHAZAVI, M.; BARARI, H. Compatibility of the entomopathogenic fungus *Beauveria bassiana* with the insecticides fipronil, pyriproxyfen and hexaflumuron. **Journal of Entomological Society of Iran**, v. 31, n. 2, p. 29-37, 2012.

REIS, M. F.; ROCHA, C. L. M. S. C. Cytological analysis of the effect of aqueous extracts of *Lentinula edodes* and *Pleurotus ostreatoroseus* on the development cycles of *Aspergillus (=Emericella) nidulans*. **SaBios: Rev. Saùde e Biol.**, v. 9, n. 1, p. 100-107, 2014.

RIBEIRO, M. M. R.; AMARAL, K. D.; SEIDE, V. E.; SOUZA, B. M. R.; DELLA LUCIA, T. M. C. ; KASUYA, M. C. M.; SOUZA, D. J. Diversity of Fungi Associated with *Atta bisphaerica* (Hymenoptera: Formicidae): The Activity of *Aspergillus ochraceus* and *Beauveria bassiana*. **Psyche**, p. 1-6, 2012.

RICCI, M., BENiTEZ, D.; PADÎN, S.; MACEIRAS, A. **Hormigas argentinas: comportamiento, distribución y control**. Argentina: Facultad de Ciencias Agrarias y Forestales, Universidad Nacional de La Plata, 2005.

ROCKWOOD, L. L.; HUBBELL, S. P. Host-plant selection, diet diversity, and optimal foraging in a tropical leafcutting ant. **Oecologia**, v. 74, p. 55-61, 1987.

RODRIGUES, A.; CARLETTI, C. D.; BUENO, O. C.; PAGNOCCA, F. C. Leafcutting ant faecal fluid and mandibular gland secretion: effects on microfungi spore germination. **Brazilian Journal of Microbiology**, v. 39, p. 64-67, 2008.

RODRIGUES, A.; PASSARINI, M. R. Z.; FERRO, M.; NAGAMOTO, N. S.; FORTI, L. C.; BACCI JR., M.; SETTE, L. D.; PAGNOCCA, F. C. Fungal communities in the garden chamber soils of leaf-cutting ants. **Journal of Basic Microbiology**, v. 54, p. 1186-1196, 2014.

RODRIGUES, C. J. B. C.; FREITAS, M. C.; PERINOTTO, W. M. S.; SANTOS, F. S.;

PAULO, J. F.; QUINELATO, S.; CAMARGO, M. G.; ANGELO, I. C.;

BITTENCOURT, V. R. E. P. Morphological study of Beauveria bassiana isolates before and after reisolation in Rhipicephalus microplus*. **Revista Brasileira de Medicina Veterinària**, v. 38, Supl. 3, p. 54-65, 2016.

RODRiGUEZ-DEL-BOSQUE, L. A.; ARREDONDO-BERNAL, H. C. **Teoria y**

Biological Control Applications. 1. ed. Mexico: Sociedad Mexicana de Control Biológico, 2007.

RODRIGUEZ, J.; CALLE, Z.; MONTOYA-LERMA, J. Herbivory of *Atta cephalotes* (Hymenoptera: Myrmicinae) on three vegetation substrates. **Revista Colombiana de Entomologia**, v. 34, n. 2, p. 156-162, 2008.

ROMER, D.; BOLLAZZI, M.; ROCES, F. Carbon dioxide sensing in an obligate insectfungus symbiosis: CO_2 preferences of leafcutting ants to rear their mutualistic fungus. **Plos One**, v. 12, n. 4, p. 1-17, 2017.

ROSSI, G. D. **Exploring host-parasitoid interactions for the identification of molecules with biotechnological potential**. 2012. Thesis (Doctorate in Sciences) - University of Sao Paulo, School of Agriculture "Luiz de Queiroz", Piracicaba, Sao Paulo.

SABBOUR, M.; ABDEL-RAHEEM, M. Efficacy of *Beauveria brongniartii* and *Nomuraea rileyi* against the potato tuber moth, *Phthorimaea operculella* (Zeller). **The American Journal of Innovative Research and Applied Sciences**, v. 1, n. 6, p. 197202, 2015.

SALINAS, E. A.; DÎAZ, O. R. A. R. Control of the "Akeke" leaf-cutting ant *Acromyrmex Iandolti* with entomopathogenic weeds. **Investigación Agraria**, v. 13, n. 1, p. 27-32, 2011.

SALVADORI, J. R.; PEREIRA, P. R. V. S. **Integrated management of weeds in wheat and associated crops**. Passo Fundo: Embrapa Trigo, 2006.

SANCHEZ, S. E. M.; FREITAS, A. L.; ROBERTS, D. W. Detection of Entomophthorales fungi pathogenic to phytophagous insects, south of Bahia, Brazil.

Entomotropica, v. 16, n. 3, p. 203-206, 2001.

SANTORO, P. H.; NEVES, P. M. O. J.; ALEXANDRE, T. M.; ALVES, L. F. A. Interference of methodology in the results of bioassays for the selection of entomopathogenic fungi for insect control. **Pesquisa Agropecuâria Brasileira**, v. 42, n. 4, p. 483-489, 2007.

SANTOS, A.; ZANETTI, R.; ZANUNCIO, J. C.; MENDONÇA, L. A.; SOUZA- SILVA, A.; MEDEIROS, A. G. B. Efficiency and cost of *Atta* spp. (Hymenoptera: Formicidae) nest control with thermal spraying. **Cerne**, v. 13, p. 23-27, 2007a.

SANTOS, A. R.; NÙNEZ, E. D. P.Aislamiento de hongos entomopatógenos en Uruguay y su virulencia sobre Trialeurodes vaporariorum West. **Agrociencia**, v. 7, n. 2, p. 71-78, 2003.

SANTOS, A. V.; DILLON, R. J.; DILLON, V. M.; REYNOLDS, S. E.; SAMUELS, R. I. Occurrence of the antibiotic producing bacterium Burkholderia sp. in colonies of the leaf-cutting ant Atta sexdens rubropilosa. **FEMS Microbiology Letters**, v. 239, p. 319-323, 2004.

SANTOS, A. V.; OLIVEIRA, B. L.; SAMUELS, R. I. Selection of entomopathogenic fungi for use in combination with sub-lethal doses of imidacloprid: perspectives for the control of the leaf-cutting ant Atta sexdens rubropilosa Forel (Hymenoptera: Formicidae). **Mycopathologia**, v. 163, p. 233-240, 2007b.

SANTOS, F. J.; FERREIRA, J. M. S.; RIBEIRO, V. J. O.; OLIVIERA, A. C. L.; AZEVEDO, A. G. C. **Protocol for mass production of entomopathogenic fungi II -** Hirsutella thompsonii **(Fischer)**. Aracaju: Embrapa Tabuleiros Costeiros, 2009.

SANTOS, G. F. **Action of** Paecilomyces lilacinus **and** Paecilomyces farinosus **on *Boophilus microplus* teleogines**. 2003. Dissertation (Postgraduate Diploma in Fungal Biology) - Federal University of Pernambuco, Recife, Pernambuco.

SANTOS, R. S. **Leafcutter ant litter and its role in plant development**. 2016. Dissertation (Post-Graduation in Ecology and Conservation) - Federal University of Sergipe, Sao Cristóvao, Sergipe.

SANTOS, T. T.; CAZETTA, M. L. Attini ants and their interaction with microorganisms. **Revista Cientifica da FHO|UNIARARAS**, v. 4, n. 1, p. 36-44, 2016.

SCHAMNE, P. A. **Effect of additives and fishfertilchitosan® in solid media on the production of conidia of *Metarhizium anisopliae* (Metsch.) Sorokin and *Beauveria bassiana* (bals.) Vuillemin**. 2010. Dissertation (Postgraduate Diploma in Agronomy) - Midwestern State University, Guarapuava, Paranà.

SCHAPOVALOFF, M. E.; ALVES, L. F. A.; URRUTIA, M. I.; LASTRA, C. C. L. Natural occurrence of entomopathogenic fungi in soils cultivated with yerba mate (Ilex paraguariensis

St. Hil.) in Misiones, Argentina. **Revista Argentina de Microbiologia**, v. 47, n. 2, p. 138-142, 2015.

SCHOENIAN, I.; SPITELLER, M.; GHASTE, M.; WIRTH, R.; HERZ, H.; SPITELLER, D. Chemical basis of the synergism and antagonism in microbial communities in the nests of leaf-cutting ants. **PNAS**, v. 108, n. 5, p. 1955-1960, 2011.

SEYE, F.; BAWIN, T.; BOUKRAA, S.; ZIMMER, J.; NDIAYE, M.; DELVIGNE, F.; FRANCIS, F. Pathogenicity of Aspergillus clavatus produced in a fungal biofilm bioreactor towards Culex quinquefasciatus (Diptera: Culicidae). **Journal of Pesticide Science**, v. 39, n. 3, p. 127-132, 2014.

SIERPINSKA, A.; POPOWSKA-NOWAK, E.; BEDNAREK, A. *Beauveria brongniartii* Sacc. (Petch) against *Melolontha* spp. white grubs in forest nurseries with different soil pH. **Folia Forestalia Polonica**, v. 57, n. 4, p. 210-217, 2015.

SILVA, C. A. D. Selection of isolates of *Beauveria bassiana* pathogenic to the cotton bollworm. **Pesquisa Agropecuâria Brasileira**, v. 36, n. 2, p. 243-247, 2001.

SILVA, C. C. A. Aspects of the insect immune system. **Biotecnologia Ciência & Desenvolvimento**, n. 24, p. 68-72, 2002.

SILVA, D. M. **Identification of species of *Aspergillus* section Nigri by polyphasic taxonomy and description of two new species of the genus**. 2009. Dissertation (Postgraduate Diploma in Agricultural Microbiology) - Federal University of Lavras, Lavras, Minas Gerais.

SILVA, G. M. H. **Algaroba fiber: an alternative substrate for the production of conidia and hydrolytic enzymes by filamentous fungi**. 2017. Monograph (Biotechnology Course) - Federal University of Paraiba, Joao Pessoa, Paraiba.

SILVA, J. C.; CASTRO, V. R.; XAVIER, B. A. **Cartilha do fazendeiro florestal**. 2. ed. Viçosa, Minas Gerais, 2008.

SILVA, L. C.; MOREIRA, R. A.; ROCHA, O. Acute toxicity of the pesticide fipronil to the cladoceran *Bosmina_ freyi* De Melo and Hebert, 1994. **Electronic Journal Fòrum Ambiental da Alta Paulista**, v. 8, n. 2, 2012.

SILVA, M. N. ***Beauveria bassiana* (BALS) Vuillemin and *Beauveria brongniartii* (Sacc) Petch: cytological characterization, proteolytic activity and morphological analysis of**

the infection process in *Boophilus microplus* and *Zabrotes subfasciatus*. 2013. Monograph (Degree in Biological Sciences) - State University of Bahia, Paulo Afonso, Bahia.

SILVA, P. S. D.; BIEBER, A. G. D.; KNOCH, T. A.; TABARELLI, M.; LEAL, I. R.; WIRTH, R. Foraging in highly dynamic environments: leaf-cutting ants adjust foraging trail networks to pioneer plant availability. **Entomologia Experimentalis et Applicata**, v. 147, p. 110-119, 2013a.

SILVA, R. A.; QUINTELA, E. D.; MASCARIN, G. M.; BARRIGOSSI, J. A. F.; LIÂO, L. M. Compatibility of conventional agrochemicals used in rice crops with the entomopathogenic fungus Metarhizium anisopliae. **Scientia Agricola**, v. 70, n. 3, p. 152-160, 2013b.

SILVA, R. Z.; NEVES, P. M. O. J. Techniques and parameters used in compatibility tests between Beauveria bassiana (Bals) Vuill and in vitro phytosanitary products. **Pest Management Science**, v. 61, p. 667-674, 2005.

SILVA, R. Z.; NEVES, P. M. O. J.; SANTORO, P. H. Techniques and parameters used in compatibility studies between entomopathogenic fungi and phytosanitary products. **Semina: Ciências Agrârias**, v. 26, n. 3, p. 305-312, 2005.

SIMAS, V. R.; COSTA, E. C.; SIMAS, C. A. Control of Camponotus punctulatus Mayr, 1868 (Hymenoptera: Formicidae). **Revista da FZVA**, v. 7/8, n. 1, p. 24-32, 2000/2001.

SIMONATO, J.; GRIGOLLI, J. F. J.; OLIVEIRA, H. N. **Biological control of insect pests in soybeans**. In: LOURENÇÂO, A. L. F.; GRIGOLLI, J. F. J.; MELOTTO, A. M.; PITOL, C.; GITTI, D. de C.; ROSCOE, R. Maracaju, MS: Fundaçao MS, 2014, p. 178-193.

SOARES, I. M. F.; DELLA LUCIA, T. M. C.; SANTOS, A. A.; NASCIMENTO, I. C.; DELABIE, J. H. C. Characterization of nests and colony size of Acromyrmex rugosus (F. Smith) (Hymenoptera, Formicidae, Attini) in sandbanks of Ilhéus, BA, Brazil. **Revista Brasileira de Entomologia**, v. 50, n. 1, p. 128-130, 2006.

SOUZA, D. J.; DELLA LUCIA, T. M. C.; ERRARD, C.; D'ETTORRE, P.; MERCIER, J. L. Recognition of offspring by workers who are nest mates and non-nest mates in Acromyrmex laticeps nigrosetosus Forel, 1908 (Hymenoptera, Formicidae). **Ciência**

Rural, v. 33, n. 1, p. 91-95, 2003.

SOUZA, R. K. F. **Hypoxia during mycelial growth induces increased tolerance of conidia of entomopathogenic fungi to heat and menadione**.

2013. Dissertation (Post-graduation in Biological Sciences) - University of Vale do Paraiba, Sao José dos Campos, Sao Paulo.

STERLING, A.; GÓMEZ, C. A.; CAMPO, A. A. Pathogenicity of Metarhizium anisopliae (Deuteromycota: Hyphomycetes) on Heterotermes tenuis (Isoptera: Rhinotermitidae) in *Hevea brasiliensis*. **Revista Colombiana de Entomologia**, v. 37, n. 1, p. 36-42, 2011.

STERN, V. M.; SMITH, R. F.; VAN DEN BOSCH, R.; HAGEN, K. S. The integrated control concept. **Hilgardia**, v. 29, n. 2, 1959.

SUEN, G.; TEILING, C.; LI, L.; HOLT, C.; ABOUHEIF, E.; BORNBERG-BAUER, E.; BOUFFARD, P.; CALDERA, E. J.; CASH, E.; CAVANAUGH, A.; DENAS, O.; ELHAIK, E.; FAVÉ, M. J.; GADAU, J.; GIBSON, J. D.; GRAUR, D.; GRUBBS, K. J.; HAGEN, D. E.; HARKINS, T. T.; HELMKAMPF, M; HU, H.; JOHNSON, B. R.; KIM, J.; MARSH, S. E.; MOELLER, J. A.; MUNOZ-TORRES, M. C.; MURPHY, M. C.; NAUGHTON, M. C.; NIGAM, S.; OVERSON, R.; RAJAKUMAR, R.; REESE, J. T.; SCOTT, J. J.; SMITH, C. R.; TAO, S.; TSUTSUI, N. D.;

VILJAKAINEN, L.; WISSLER, L.; YANDELL, M. D.; ZIMMER, F.; TAYLOR, J.; SLATER, S. C.; CLIFTON, S. W.; WARREN, W. C.; ELSIK, C. G.; SMITH, C. D.; WEINSTOCK, G. M.; GERARDO, N. M.; CURRIE, C. R. The Genome Sequence of the Leaf-Cutter Ant *Atta cephalotes* Reveals Insights into Its Obligate Symbiotic Lifestyle. **PLOS**, v. 7, n. 2, p. 1-11, 2011.

SVEDESE, V. M. **Infection process of *Beauveria bassiana* on the sugarcane borer *Diatraea saccharalis* (Lepidoptera: Crambidae)**. 2012. Thesis (Post-graduation in Biological Sciences) - Federal University of Pernambuco. Recife, Pernambuco.

TAMAI, M. A.; ALVES, S. B.; ALMEIDA, J. E. M.; FAION, M. Evaluation of entomopathogenic fungi for the control of *Tetranychus urticae* Koch (Acari: Tetranychidae). **Arquivos do Instituto Biológico**, v. 69, n. 3, p. 77-84, 2002a.

TAMAI, M. A.; ALVES, S. B.; LOPES, R. B.; FAION, M.; PADULLA, L. F. L. Toxicity of phytosanitary products to *Beauveria bassiana* (Bals.) Vuill. **Arquivos do Instituto**

Biológico, v. 69, n. 3, p. 89-96, 2002b.

TAMAI, M. A.; ALVES, S. B.; NEVES, P. J. Pathogenicity of *Beauveria bassiana* (Bals.) Vuill. to the mite *Tetranychus urticae* KOCH. **Scientia Agricola**, v. 56, n. 2, 1999.

TANADA, Y.; KAYA, H. K. **Insect Pathology**. Academic Press, London, 1993.

TANZINI, M. R.; ALVES, S. B.; SETTEN, A. Toxicity of phytosanitary products used to control *Leptopharsa heveae* to entomopathogenic fungi.

Archives of the Biological Institute, v. 69, n. 4, p. 65-69, 2002.

TAVARES, M. M. **O papel do ar e à água ambiental como veiculos de transmissâo de Infeções fùngicas no Hospital Agostinho Neto, Cidade da Praia, Cabo Verde**.

2012. Dissertation (Master's Degree in Molecular Genetics) - University of Minho, Portugal.

TEIXEIRA, G. T.; JAFELICE, R. S. M.; ALMEIDA, C. G.; VASCONCELOS, H. L. Cellular automata in the study of the reoccupation of leafcutter ants in foraging regions. **Biomathematics**, v. 24, p. 77-90, 2014.

TEJEDA, C. S.; GARCiA, A. A.; TORRES, B. C. P.; OLGUÌN, J. F. L. Alternativa Agroecológica para el Manejo de Atta mexicana en Puebla, México. **Southwestern Entomologist**, v. 42, n. 1, p. 261-273, 2017.

TÉLLEZ-JURADO, A.; RAMiREZ, M. G. C.; FLORES, Y. M.; TORRES, A. A.; ARANA-CUENCA, A. Mecanismos de acción y respuesta en la relación de hongos entomopatógenos y insectos. **Revista Mexicana de Micologia**, v. 30, p.73-80, 2009.

THOMPSON, J. J. W.; ARMITAGE, S. A. O.; SIVA-JOTHY, M. T. Cuticular color change after imaginal eclosion is time-constrained: blacker beetles darken faster.

Physiological Entomology, v. 27, p. 136-141, 2002.

TIAGO, P. V.; FURLANETO, M. C. The role of cuticle-degrading proteases produced by entomopathogenic fungi. **Revista do Programa de Ciências Agro- Ambientais**, v. 2, n. 1, p. 40-51, 2003.

TOFOLO, V. C.; GIANNOTTI, E.; PIZANO, M. A. Exposure of Workers of Ectatomma brunneum Smith (Hymenoptera: Formicidae: Ectatomminae) to ant Baits Containing Different Active Ingredients under Laboratory Conditions.

EntomoBrasilis, v. 8, n. 1, p. 38-44, 2015.

TORRES, A. F.; LASMAR, O.; CARVALHO, G. A.; SANTA-CECILIA, L. V. C.; ZANETTI, R.; OLIVEIRA, D. Insecticidal activity of plant extracts in the control of leaf-cutting ants in coffee trees. **Coffee Science**, v. 8, n. 3, p. 371-378, 2013.

TRANTER, C.; HUGHES, W. O. H. Acid, silk and grooming: alternative strategies in social immunity in ants? **Behavioral Ecology and Sociobiology**, v. 69, p. 1687-1699, 2015.

TRAVAGLINI, R. V.; STEFANELLI, L. E. P.; ARNOSTI, A.; CAMARGO, R. S.; FORTI, L. C. Attractive encapsulated bait for microbial control of leaf-cutting ants. **Tekhne e Logos**, v. 8, n. 3, p. 100-111, 2017.

UDO, M. S. B. **Evaluation of the toxic effects of prenatal exposure to Fipronil on the offspring of Wistar rats**. 2012. Dissertation (Master's Degree) - University of Sao Paulo, Sao Paulo.

VALICENTE, F. H. Biological control of pests with entomopathogens. **Informe Agropecuârio**, v. 30, n. 251, p. 48-55, 2009.

VAN DEN BOSCH, R.; MESSENGER, P. S.; GUITIEREEZ, A. P. **An introduction to biological control**. New York: Plenum Press, 1982.

VEGA, F. E.; JACKSON, M. A.; MCGUIRE, M. R. Germination of conidia and blastospores of Paecilomyces fumosoroseus on the cuticle of the silverleaf whitefly, Bemisia argentifolii. **Mycopathologia**, v. 147, p. 33-35, 1999.

VELASQUEZ, V. B.; CARCAMO, M. P.; MERINO, C. R.; IGLESIAS, A. F.; DURAN, J. F. Intraspecific differentiation of Chilean isolates of the entomopathogenic fungi Metarhizium anisopliae var. anisopliae as revealed by RAPD, SSR and ITS markers. **Genetics and Molecular Biology**, v. 30, n. 1, p. 8999, 2007.

VERiSSIMO, C. J. **Controle de tickatos nas pastagens**. 2. ed. Nova Odessa, Sao Paulo: Instituto de Zootecnia, 2015.

VERiSSIMO, C. M. N. **Invasive aspergillosis in bone marrow transplant patients**. 2007. Dissertation (Master's Degree in Human Biology and Environment) - University of Lisbon, Portugal.

VIGUERAS, G.; PAREDES-HERNANDEZ, D.; REVAH, S.; VALENZUELA, J.; OLIVARES-HERNANDEZ, R.; LE BORGNE, S. Growth and enzymatic activity of

Leucoagaricus gongylophorus, a mutualistic fungus isolated from the leaf-cutting ant Atta mexicana, on cellulose and lignocellulosic biomass. **Letters in Applied Microbiology**, v. 65, p. 173-181, 2017.

VILELA, E. F. **Status of leaf-cutting and control in forest plantations in Brazil.**

In: LOFGREN, C. S.; VANDERMEER, R. K. Fire ants and leafcutting ants: biology and management. Boulder: Wesview Press, 1986.

VITTAR, F. Hormigas (Hymenoptera: Formicidae) de la Mesopotamia Argentina. **INSUGEO**, Miscelânea, v. 17, n. 2, p. 447-466, 2008.

WENZEL, I. M. **Pathogenicity of *Lecanicillium lecanii* (ZIMM.) ZARE & GAMS to the striped mite *Tetranychus urticae* (ACARI: TETRANYCHIDAE) and its Compatibility with agrochemicals and biocontrol organisms utilized in the**

chrysanthemum culture. 2005. Thesis (Doctorate in Agronomy) - Universidade Estadual Paulista "Jùlio de Mesquita Filho", Botucatu, Sao Paulo.

WILSON, K.; COTTER, S. C. **Density-dependent prophylaxis in insects**. p. 137176. In: ANANTHAKRISHNAN, T. N.; WHITMAN, D.W. Insects and Phenotypic Plasticity: Mechanisms and Consequences. Plymouth, UK: Science Publishers Inc.

XAVIER, L. M. S.; AVILA, C. J. Pathogenicity, LD50 and TL50 of isolates of *Metarhizium anisopliae* (Metsch.) Sorok. for the brown root bug *Scaptocoris carvalhoi* Becker (Hemiptera: Cydnidae). **Ciência Rural**, v. 35, n. 4, p. 763-768, 2005.

XAVIER, M. O.; MADRID, I. M.; CLEFF, M. B.; CABANA, A. L.; FILHO, R. P.

S.; MEIRELES, M. C. A. Air contamination by *Aspergillus* in a marine animal rehabilitation environment. **Brazilian Journal of Veterinary Research and Animal Science**, v. 45, n. 3, p. 174-179, 2008.

XIAO, G.; YING, S.; ZHENG, P.; WANG, Z.; ZHANG, S.; XIE, X.; SHANG, Y.; ST LEGER, R. J.; ZHAO, G.; WANG, C.; FENG, M. Genomic perspectives on the evolution of fungal entomopathogenicity in *Beauveria bassiana*. **Scientific Reports**, v. 2, p. 1-10, 2012.

YÂNEZ, M.; FRANCE, A. Effects of fungicides on the development of the entomopathogenic fungus *Metarhizium anisopliae* var. *anisopliae*. **Chilean Journal of Agricultural Research**, v. 70, n. 3, p. 390-398, 2010.

ZANETTI, R.; ZANUNCIO, J. C.; SANTOS, J. C.; SILVA, W. L. P.; RIBEIRO, G.; T.; LEMES, P. G. An Overview of Integrated Management of Leaf-Cutting Ants (Hymenoptera: Formicidae) in Brazilian Forest Plantations. **Forests**, v. 5, p. 439-454, 2014.

ZANETTI, R.; ZANUNCIO, J. C.; SOUZA-SILVA, A.; MENDONÇA, L. A.;

MATTOS, J. O. S.; RIZENTAL, M. S. Efficiency of thermonebuligenic products in controlling *Atta*

laevigata (Hymenoptera: Formicidae) in eucalyptus plantations. **Ciência e Agrotecnologia**, v. 32, n. 4, p. 1313-1316, 2008.

ZARZUELA, M. F. M. **Isolation of entomopathogens in invasive ant colonies and their application for control**. 2010. Thesis (Doctorate in Biological Sciences) - Universidade Estadual Paulista "Jùlio de Mesquita Filho", Rio Claro, Sao Paulo.

ZHANG, Y.; ZHANG, J.; JIANG, X.; WANG, G.; LUO, Z.; FAN, Y.; WU, Z.; PEI, Y. Requirement of a Mitogen-Activated Protein Kinase for Appressorium Formation

MIX
Papier aus verantwortungsvollen Quellen
Paper from responsible sources
FSC® C105338

Printed by Books on Demand GmbH, Norderstedt / Germany